T0192408

Analytic Methods for Diophantine Equations
and Diophantine Inequalities

Harold Davenport was one of the truly great mathematicians of the twentieth century. Based on lectures he gave at the University of Michigan in the early 1960s, this book is concerned with the use of analytic methods in the study of integer solutions to Diophantine equations and Diophantine inequalities. It provides an excellent introduction to a timeless area of number theory that is still as widely researched today as it was when the book originally appeared. The three main themes of the book are Waring's problem and the representation of integers by diagonal forms, the solubility in integers of systems of forms in many variables, and the solubility in integers of diagonal inequalities.

For the second edition of the book a comprehensive foreword has been added in which three leading experts describe the modern context and recent developments. A complete list of references has also been added.

# Analytic Methods for Diophantine Equations and Diophantine Inequalities

*Second edition*

H. Davenport

*This edition edited and prepared for publication by*

T. D. Browning
*Oxford University*

**CAMBRIDGE**
**UNIVERSITY PRESS**

University Printing House, Cambridge CB2 8BS, United Kingdom

One Liberty Plaza, 20th Floor, New York, NY 10006, USA

477 Williamstown Road, Port Melbourne, VIC 3207, Australia

314-321, 3rd Floor, Plot 3, Splendor Forum, Jasola District Centre, New Delhi - 110025, India

79 Anson Road, #06-04/06, Singapore 079906

Cambridge University Press is part of the University of Cambridge.

It furthers the University's mission by disseminating knowledge in the pursuit of education, learning and research at the highest international levels of excellence.

www.cambridge.org
Information on this title: www.cambridge.org/9780521605830



First edition first published by Campus Publishers 1963
This edition first published 2005

*A catalogue record for this publication is available from the British Library*

ISBN 978-0-521-60583-0 Paperback

# Contents

# Foreword

### Waring's problem: Chapters 1–10

When Davenport produced these lecture notes there had been very little progress on Waring's problem since important work by Davenport and Vinogradov something like a quarter of a century earlier, and the main interest was to report on the more recent work on forms as described in the later chapters. Indeed there was a generally held view, with regard to Waring's problem at least, that they had extracted everything that could be obtained reasonably by the Hardy–Littlewood method and that the method was largely played out. Moreover, the material on Waring's problem was not intended, in general, to be state of the art, but rather simply an introduction to the Hardy–Littlewood method, with a minimum of fuss by a masterly expositor, which could then be developed as necessary for use in the study of the representation of zero by general integral forms, especially cubic forms, in the later chapters. There is no account of Davenport's own fundamental work on Waring's problem, namely $G(4) = 16$ (Davenport [18]), $G(5) \leq 23$, $G(6) \leq 36$ (Davenport [19]), nor of Vinogradov's [94] $G(k) \leq 2k \log k + o(k \log k)$ for large $k$ or Davenport's proof [17] that almost all natural numbers are the sum of four positive cubes. Nor, on a more technical level, was any attempt made to obtain more refined versions of Lemmas 4.2 and 9.2, estimates for the generating function $T(\alpha)$ on the major arcs, such as those due to Davenport and Heilbronn [25] or Hua [50], although such refinements can be very helpful in applications.

In the last twenty years there has been a good deal of progress on Waring's problem. Methods of great flexibility, inspired by some of the ideas stemming from the researches of Hardy and Littlewood, Davenport, and Vinogradov have been developed which have permitted the retention of many of the wrinkles introduced in the earlier methods. The beginnings of a glimmer of some of these seminal ideas can be seen in Lemmas 9.4 and 9.5.

The asymptotic formula for the number of representations of a large natural number $n$ as the sum of at most $s$ $k$th powers established in Theorem 4.1 when $s \geq 2^k + 1$ was state of the art for $3 \leq k \leq 10$, but for larger $k$ methods due to Vinogradov were superior (see Theorem 5.4 of Vaughan [86]). The current state of play is that the asymptotic formula is known to hold when $s \geq 2^k$ $(k = 3, 4, 5)$ (Vaughan [82, 84]), $s \geq 7.2^{k-3}$ $(k = 6, 7, 8)$ (Boklan [8], following Heath-Brown [43]), and $s \geq s_1(k)$ where $s_1(k) = k^2\big(\log k + \log\log k + O(1)\big)$ $(k \geq 9)$ (Ford [32]). The discussion in the Note in Chapter 3 in the case $k = 3$ is still relevant today. Although the asymptotic formula for sums of eight cubes is now established the classical convexity bound was not improved in the exponent when $2 < m < 4$. The core of the argument of Vaughan [82] is extremely delicate and leads only to

$$\int_{\mathfrak{m}} |T(\alpha)|^8 d\alpha \ll P^5 (\log P)^{-\gamma}$$

for a positive constant $\gamma$ and a suitable set of minor arcs $\mathfrak{m}$. However Hooley [47] has shown under the (unproven) Riemann Hypothesis for a certain Hasse–Weil $L$-function that

$$\int_{\mathfrak{m}} |T(\alpha)|^6 d\alpha \ll P^{3+\varepsilon}$$

and this in turn implies the asymptotic formula for sums of seven cubes. Unfortunately it is not even known whether the $L$-function has an analytic continuation into the critical strip.

For $G(k)$ the best results that we currently have are $G(3) \leq 7$ (Linnik [57, 59]), $G(4) = 16$ Davenport [18], $G(5) \leq 17$, $G(7) \leq 33$, (Vaughan and Wooley [89]), $G(6) \leq 21$ (Vaughan and Wooley [88]), $G(8) \leq 42$ (Vaughan and Wooley [87]), $G(9) \leq 50$, $G(10) \leq 59$, $G(11) \leq 67$, $G(12) \leq 76$, $G(13) \leq 84$, $G(14) \leq 92$, $G(15) \leq 100$, $G(16) \leq 109$, $G(17) \leq 117$, $G(18) \leq 125$, $G(19) \leq 134$, $G(20) \leq 142$ (Vaughan and Wooley [90]), and $G(k) \leq s_2(k)$ where $s_2(k) = k\big(\log k + \log\log k + O(1)\big)$ (Wooley [98]) in general. Let $G^{\#}(4)$ denote the smallest positive $s$ such that whenever $1 \leq r \leq s$ every sufficiently large $n$ in the residue class $r$ modulo 16 is the sum of at most $s$ fourth powers. Then, in fact, Davenport showed that $G^{\#}(4) \leq 14$ and we now can prove (Vaughan [85]) that $G^{\#}(4) \leq 12$. Linnik's work on Waring's problem for cubes does not use the Hardy–Littlewood method, but instead is based on the theory of ternary quadratic forms. Watson [95] gave a similar but simpler proof. However these proofs give relatively poor information about the number of representations as a sum of seven cubes. As part of the recent progress we now have proofs via the Hardy–Littlewood method (e.g. Vaughan [85]) which give lower bounds of the expected correct order of magnitude for the number of representations. Davenport gives no indication of what he might have believed the correct value of $G(k)$ to be. The simplest guess is that

$$G(k) = \max\{k+1, \Gamma(k)\}$$

where $\Gamma(k)$ is as defined in the paragraph just prior to Theorem 5.1. This would imply that for $k \geq 3$, $G(k) = 4k$ when $k = 2^l$ and $k+1 \leq G(k) \leq \frac{3}{2}k$ when $k \neq 2^l$.

With regard to Lemma 9.2 and the Note after the proof, we now know that under the less stringent hypothesis $(q,a) = 1$, $q|\beta| \leq \frac{1}{2k}P^{1-k}$, $\alpha = \beta + a/q$ we have the stronger estimate

$$T(\alpha) = q^{-1}S_{a,q}I(\beta) + O\big(q^{\frac{1}{2}+\varepsilon}\big).$$

Moreover with only the hypothesis $(q,a) = 1$ we have

$$T(\alpha) = q^{-1}S_{a,q}I(\beta) + O\big(q^{\frac{1}{2}+\varepsilon}(1+P^k|\beta|)^{\frac{1}{2}}\big).$$

See Theorem 4.1 of Vaughan [86]. The latter result enables a treatment to be given for cubes in which all the arcs are major arcs.

For a modern introduction to the Hardy–Littlewood method and some of the more recent developments as applied to Waring's problem see Vaughan [86], and for a comprehensive survey of Waring's problem see Vaughan and Wooley [91].

Chapter 7 is concerned with the solubility, given a sequence $\{c_j\}$ of natural numbers, of the equation

$$c_1x_1^k + \cdots + c_sx_s^k = N \tag{1}$$

for large natural numbers $N$, and is really a warm-up for Chapters 8 and 10. For an infinite set of $N$ there may not be solutions, however large one takes $s$ to be, but the obstruction is purely a local one. Any of the various forms of the Hardy–Littlewood method which have been developed for treating Waring's problem are readily adjusted to this slightly more general situation and, with the corresponding condition on $s$, lead to an approximate formula for the number of solutions counted. This will lead to a positive lower bound for the number of solutions for any large $N$ for which the singular series is bounded away from 0.

Davenport gives a brief outline of the minor changes in the argument which have to be made in adapting the method, and the remainder of the chapter is devoted to showing that the above condition on the singular series is essentially equivalent to the expected local solubility condition.

In Chapters 8 and 10, Davenport adapts the method to treat

$$c_1 x_1^k + \cdots + c_s x_s^k = 0 \tag{2}$$

where now the $c_j$ can be integers, and not all the same sign when $k$ is even. Of course this has a solution, and so the main point of interest is to establish the existence of integral solutions in which not all the $x_j$ are 0. This can be considered to be the first special case of what was the main concern of these notes, namely to investigate the non-trivial representation of 0 by general forms and systems of forms. In Chapter 8 the simplest version of the Hardy–Littlewood method developed in the previous chapters is suitably adapted. This requires quite a large value of $s$ to ensure a solution. In Chapter 10 this requirement is relaxed somewhat by adapting the variant of Vinogradov's argument used to treat Waring's problem in Chapter 9. Although the argument of Chapter 10 is relatively simple it is flawed from a philosophical point of view in that as well as the local solubility of (2) there needs to be a discussion of the local solubility of (1) with $N$ non-zero, which, of course, really should not be necessary. This could have been avoided, albeit with some complications of detail. The question of the size of $s$ to ensure a non-trivial solution to (2) had some independent interest as Davenport and Lewis [27] had shown that $k^2+1$ variables suffice for the singular series to be bounded away from 0, and when $k+1$ is prime there are equations in $k^2$ variables with no non-trivial solution. Moreover they had also shown, via the Hardy–Littlewood method, that (2) is soluble when $s \geq k^2+1$ and either $k \leq 6$ or $k \geq 18$. Later in Vaughan [81] ($11 \leq k \leq 17$), [83] ($7 \leq k \leq 9$) and [85] ($k = 10$) this gap was removed. The methods of Vaughan and Wooley mentioned in connection with Waring's problem when adapted show that far fewer variables suffice for a non-trivial solution to (2) provided that the corresponding singular series is bounded away from 0, and this is essentially equivalent to a local solubility condition.

In the later chapters the Hardy–Littlewood method is adapted in various, sometimes quite sophisticated, ways. However, the only place where any of the main results of the first 10 chapters is applied directly is the use of Theorem 8.1 (or Theorem 10.1) in the proof of Birch's theorem in Chapter 11. Later Birch [7] gave a completely elementary proof, based partly on methods of Linnik [58], of a result similar to Theorem 8.1 which can be used in its place.

R. C. Vaughan
*Pennsylvania State University*

### Forms in many variables: Chapters 11–19

Let $F(x_1, \ldots, x_n)$ be a form of degree $d$ with integer coefficients. When $d \geq 3$, the question of whether the equation $F(x_1, \ldots, x_n) = 0$ has a non-trivial integer solution is extremely natural, extremely general, and extremely hard. However for quadratic forms a complete answer is given by the Hasse–Minkowski Theorem, which states that there is a non-trivial solution if and only if there is such a solution in $\mathbb{R}$ and in each $p$-adic field $\mathbb{Q}_p$. Such a result is known to be false for higher degree forms, as Selmer's example

$$3x_1^3 + 4x_2^3 + 5x_3^3 = 0$$

shows. None the less the hope remains that if the number of variables is not too small we should still have a 'local-to-global' principle, of the type given by the Hasse–Minkowski Theorem.

It transpires that the $p$-adic condition holds automatically if the number of variables $n$ is sufficiently large in terms of the degree. This was shown by Brauer [9], whose argument constitutes the first general method for such problems. The line of attack uses multiply nested inductions, and in consequence the necessary number of variables is very large. It was conjectured by Artin that $d^2 + 1$ variables always suffice, there being easy examples of forms in $d^2$ variables with only trivial $p$-adic solutions. However many counter-examples have subsequently been discovered. The first of these, due to Terjanian [80], involves a quartic form in 18 variables, with no non-trivial 2-adic solution. There are no known counter-examples involving forms of prime degree, and in this case it remains an open question whether or not Artin's conjecture holds.

There are various alternatives to Brauer's induction approach for the $p$-adic problem. Davenport presents one of these for the case $d = 3$ in Chapter 18, establishing the best possible result, namely that $p$-adic solutions always exist when $n \geq 10$. For $d \geq 4$ such approaches work well only when $p$ is large enough. Thus Leep and Yeomans [55] have shown that $p \geq 47$ suffices for $d = 5$. In the general case Ax and Kochen [1] showed that $d^2 + 1$ variables always suffice for the $p$-adic problem, when $p$ is sufficiently large compared with $d$. The Ax–Kochen proof is remarkable for its use of methods from mathematical logic. For small primes other lines of argument seem to be needed, and Wooley [100] has re-visited the Brauer induction approach to establish that $d^{2^d}+1$ variables suffice for every field $\mathbb{Q}_p$. It remains a significant open problem to get bounds of a reasonable size, below 1000 say, for the cases $d=4$ and $d=5$.

The problem for forms over $\mathbb{Q}$, rather than $\mathbb{Q}_p$, is distinctly different. For forms of even degree there is no value of $n$ which will ensure the existence of a non-trivial integer solution, as the example

$$x_1^d+\cdots+x_n^d=0$$

shows. Thus the original Brauer induction argument cannot be applied to $\mathbb{Q}$, since it involves an induction over the degree. However Birch [5] was able to adapt the induction approach so as to use forms of odd degree only, and hence to show that for any odd integer $d\geq 1$ there is a corresponding $n(d)$ such that $F(x_1,\ldots,x_n)=0$ always has a non-trivial solution for $n\geq n(d)$. This work is described by Davenport in Chapter 11. A rather slicker account is now available in the book by Vaughan [86, Chapter 9]. Although the values of $n(d)$ produced by Birch's work were too large to write down, more reasonable estimates have been provided by Wooley [99], by a careful adaptation of Birch's approach.

Davenport's own major contribution to the area was his attack on cubic forms, via the circle method. The natural application of Weyl's method, as described in Chapter 13, leads to a system of Diophantine inequalities involving bilinear forms. The key result in this context is Lemma 13.2. By using techniques from the geometry of numbers, Davenport was able to convert these inequalities into equations. In his first two papers on the subject [20, 21] these equations were used to deduce that $F$ must represent a form of the type $a_1x_1^3+F'(x_2,\ldots,x_m)$ for some $m<n$. This process is somewhat wasteful, since $n-m$ variables are effectively discarded. By repeated applications of the above principle Davenport was able to reduce consideration to diagonal forms. Davenport's third paper [22] treats the bilinear equations in a more geometrical way, which is presented in Chapter 14. This approach is much more efficient, since no variables are wasted. A straightforward application of this third method shows that $F=0$ has a non-trivial solution for any cubic form in 17 or more variables, and this is the result given as Theorem 18.1. However in [22] a slight refinement is used to show that 16 variables suffice. It is perhaps worth emphasizing the slightly unusual logical

structure of the proof. The main goal is to prove an asymptotic formula for the number of solutions in a box of side $P$. Davenport achieves this, providing that the number of solutions to the aforementioned bilinear equations does not grow too rapidly. The arguments used to handle this latter issue lead to two alternatives: either the number of solutions to the bilinear equations is indeed suitably bounded, or the original cubic form has a non-trivial integer zero for geometric reasons. In either case the cubic form has a non-trivial integer zero. One consequence of all this is that one does not obtain an asymptotic formula in every case. The form

$$x_1^3 + x_2(x_3^2 + \cdots + x_n^2)$$

vanishes whenever $x_1 = x_2 = 0$, so that there are $\gg P^{n-2}$ solutions in a box of side $P$. This example shows that one cannot in general expect an asymptotic formula of the type mentioned in connection with Theorem 17.1.

The 16 variable result is arguably one of Davenport's finest achievements, and it remains an important challenge to show that 15 variables, say, are in fact enough. Davenport's approach has been vastly generalized by Schmidt [77] so as to apply to general systems of forms of arbitrary degree. For a single form $F(x_1, \ldots, x_n)$ the result may be expressed in terms of the invariant $h(F)$ defined as the smallest integer $h$ for which one can write

$$F(\mathbf{x}) = G_1(\mathbf{x})H_1(\mathbf{x}) + \cdots + G_h(\mathbf{x})H_h(\mathbf{x})$$

with non-constant forms $G_i, H_i$ having rational coefficients. An inspection of Davenport's argument for cubic forms in 16 variables then establishes the standard Hardy–Littlewood asymptotic formula for any cubic form with $h(F) \geq 16$. When $h(F) \leq 15$ and $n \geq 16$ the form $F$ still has a non-trivial integer zero, since one can take the forms $H_i(\mathbf{x})$ to be linear and use a common zero of $H_1, \ldots, H_h$. In his generalization Schmidt was able to obtain an explicit function $n(d)$ such that the Hardy–Littlewood formula holds for any form of degree $d$ having $h(F) \geq n(d)$. In order to deal with forms for which $h(F) < n(d)$ one is led to an induction argument involving systems of forms. Thus if one starts with a single form of degree $d = 5$ one wants to know about zeros of systems of cubic forms. In this connection Schmidt proved in a separate investigation [76] that a system of $r$ cubic forms with integer coefficients has a non-trivial integer zero if there are at least $(10r)^5$ variables.

Davenport's result was generalized in another direction by Pleasants

[67], who showed that the result remains true if the coefficients of the form $F$, and the solutions $(x_1, \ldots, x_n)$, are allowed to lie in an algebraic number field. In this wider setting 16 variables still suffice.

If one assumes the form $F$ to be non-singular, which is the generic case, one can show (Heath-Brown [42]) that 10 variables suffice. Here the number 10 is best possible, since there exist forms in 9 variables with no non-trivial $p$-adic zeros. However Hooley [45] has sharpened the above result to establish the local-to-global principle for non-singular cubic forms in $n \geq 9$ variables. These works use the Hardy–Littlewood method, but instead of employing Weyl's inequality they depend on the Poisson summation formula and estimates for 'complete' exponential sums. Complete exponential sums involving a non-singular form can be estimated very efficiently via Deligne's Riemann Hypothesis for varieties over finite fields, but the methods become less effective as the dimension of the singular locus grows. Deligne's bounds handle sums to prime, or square-free, moduli, but sums to prime power moduli remain a considerable problem. The treatment of these in [42] uses exactly the same bilinear forms as were encountered by Davenport [22], but since $F$ is now non-singular the techniques of Birch [6] can be used to advantage. Heath-Brown [42] establishes an asymptotic formula for the number of solutions in a suitable region. However the argument in Hooley [45] has a structure somewhat analogous to Davenport's, in that one only gets an asymptotic formula under a certain geometric condition. When the condition fails there are integer points for other reasons. (This defect was later circumvented by Hooley [46].) In its simplest guise the above methods would handle non-singular cubic forms in 13 or more variables. However this may be reduced to 10 through the use of Kloosterman's refinement of the circle method. In order to handle forms in nine variables Hooley adopts a distinctly more subtle analysis, designed to save just a power of $\log P$, when considering points in a box of side $P$.

The work of Birch [6], summarized in Chapter 19, is most easily described by seeing how it applies to a single form $F$. When $F$ is non-singular Birch is able to establish an asymptotic formula as soon as $n > (d-1)2^d$, providing that the singular series and integral are positive. For $d = 3$ this is weaker than the result of Hooley [45], but the method works for arbitrary values of $d$. In fact subsequent investigations have failed to improve on Birch's result for any value of $d > 3$. Birch's argument is based on Weyl's inequality, and leads to a system of multilinear equations analogous to the bilinear ones in Davenport's work. These are handled by a different technique from that used by

Davenport, which is simpler and more obviously geometric, but which requires information about the singularities of $F$.

D. R. Heath-Brown
*Oxford University*

## Diophantine inequalities: Chapter 20

In the final chapter, Davenport provides an exposition of his ground-breaking 1946 joint work with Heilbronn [26]. They demonstrated how to adapt the Hardy–Littlewood method to yield results on Diophantine inequalities. Since their publication, numerous results have been proved with their technique, now commonly referred to as the Davenport–Heilbronn method.

Suppose that $s$ is an integer with $s \geq 5$ and that $\lambda_1, \ldots, \lambda_s$ are real numbers, not all of the same sign, and not all in rational ratio. The chapter consists of a proof that given any positive real number $C$, there exists a non-trivial integer solution $\mathbf{x} = (x_1, \ldots, x_s)$ of the Diophantine inequality

$$\left|\lambda_1 x_1^2 + \cdots + \lambda_s x_s^2\right| < C. \tag{3}$$

As Davenport notes, the result has a straightforward extension to the case in which the squares are replaced by $k$th powers and the number of variables is at least $2^k + 1$. If $k$ is odd, the sign condition is of course unnecessary.

The proof is a clever adaptation of the Hardy–Littlewood method. One estimates, for some large positive $P$, the number of solutions of (3) where the integers $x_i$ satisfy $|x_i| \leq P$. Rather than integrating over a unit interval as in the Hardy–Littlewood method, one integrates over the real line against a suitable decaying kernel. Instead of multiple major arcs, here the major contribution comes from an interval centred around zero, while the most difficult region to bound consists of a subset of numbers of intermediate size. The contribution to this latter region is treated using the hypothesis that one of the ratios is irrational.

In the lecture notes, Davenport conjectures that (3) is non-trivially soluble even for $s \geq 3$, and in a separate comment notes that a natural question is whether the result can be generalized to the case of indefinite quadratic forms that are not necessarily diagonal and discusses some work by Birch, Davenport and Ridout (see [29]). In fact, Margulis [60] answered both questions positively, establishing the non-trivial solubility of

$$|Q(x_1,\ldots,x_s)| < \varepsilon$$

for general indefinite quadratic forms $Q(\mathbf{x})$, for any $\varepsilon > 0$, assuming $s \geq 3$ and that the coefficients of $Q$ are not all in rational ratio. This established the Oppenheim conjecture, as it implies that the values of such a form at integral points are dense on the real line. We note that Margulis' proof uses techniques different from the Hardy–Littlewood method.

Concerning forms of higher degree, Davenport mentions a result that Pitman [66] gave on cubic forms, but remarks that proving similar results for forms of higher odd degree seems to involve a 'difficulty of principle'. Schmidt, in a sequence of papers [73, 74, 75], provided the key result needed to resolve this difficulty. His work builds on a combination of the Davenport–Heilbronn method and a diagonalization procedure that yields a proof that any system of general Diophantine inequalities of odd degree and sufficiently many variables has a solution. More precisely, he showed that given odd positive integers $d_1,\ldots,d_R$, there exists a constant $C(d_1,\ldots,d_R)$ depending only on $d_1,\ldots,d_R$ such that given any real forms $F_1,\ldots,F_R$ in $s$ variables, of respective degrees $d_1,\ldots,d_R$, where $s \geq C(d_1,\ldots,d_R)$, and given $\varepsilon > 0$, there exists a non-trivial integer vector $\mathbf{x}$ such that

$$|F_1(\mathbf{x})| < \varepsilon, \quad |F_2(\mathbf{x})| < \varepsilon, \quad \ldots \quad , \quad |F_R(\mathbf{x})| < \varepsilon.$$

There are numerous results which give lower bounds such as $C(d_1,\ldots, d_R)$ for particular types of forms, of which we mention only two. Brüdern and Cook [11] produced such a result for systems of diagonal forms, under certain conditions on the coefficients, and Nadesalingam and Pitman [62] have given an explicit lower bound for systems of $R$ diagonal cubic forms.

One can also ask about inequalities involving general positive definite forms with coefficients not all in rational ratio. We certainly do not expect the values at integral points to be dense on the real line; thus the relevant question, asked by Estermann, is whether the gaps between these values tend to zero as the values tend to infinity, provided that the number of variables is sufficiently large. For diagonal quadratic forms, Davenport and Lewis [28] noted that this follows readily from a result of Jarník and Walfisz [51], if the number of variables $s$ is at least 5. In their paper, Davenport and Lewis gave a step toward answering the gaps question for general positive definite quadratic forms $Q(\mathbf{x})$ in $s$

variables. Their methods essentially show, as Cook and Raghavan [15] demonstrate, that for such forms, given $s$ sufficiently large and given $\varepsilon > 0$, then for any sufficiently large integral point $\mathbf{x}_0$, there are many integral points $\mathbf{x}$ for which one has $|Q(\mathbf{x}) - Q(\mathbf{x}_0)| < \varepsilon$, where the notion of many can be defined precisely. In 1999, Bentkus and Götze [3] resolved the gaps question with powerful new techniques, which Götze [36] consequently improved upon. These results together establish that for $s \geq 5$ and for any positive definite quadratic form $Q$ in $s$ variables, with coefficients not all in rational ratio, the differences between successive values of $Q$ at integral points tend to zero as the values approach infinity. Their methods have given rise to much new work on Diophantine inequalities. Additionally, we note that some workers have considered special types of inhomogeneous polynomials of higher degree, including Brüdern [10], Bentkus and Götze [4] and Freeman [34].

Since Davenport and Heilbronn's work, there have been many improvements of the lower bound on $s$ required to guarantee non-trivial solubility of diagonal Diophantine inequalities of degree $k$. For each positive integer $k$, let $G_{\text{ineq}}(k)$ denote the smallest positive integer $s_0$ such that for all $s \geq s_0$, and for all indefinite diagonal forms $\lambda_1 x_1^k + \cdots + \lambda_s x_s^k$ with coefficients not all in rational ratio, and for all $\varepsilon > 0$, there is a non-trivial integral solution of

$$\left|\lambda_1 x_1^k + \cdots + \lambda_s x_s^k\right| < \varepsilon. \tag{4}$$

As Davenport remarks, Davenport and Roth [30] provided an improvement; they showed that there exists a constant $C_1 > 0$ such that

$$G_{\text{ineq}}(k) \leq C_1 k \log k.$$

In fact, the Davenport–Heilbronn method is sufficiently flexible so that bounds for inequalities roughly parallel bounds given by work on Waring's problem. In particular, for large $k$, one has

$$G_{\text{ineq}}(k) \leq k(\log k + \log\log k + 2 + o(1)). \tag{5}$$

(See [101] for a statement of this result.) We note that in many cases, for example the work of Baker, Brüdern and Wooley [2] for $k = 3$, achieving the same bound as that for $G(k)$ required extra effort. Recent work of Wooley [101] shows that bounds for $G(k)$ generally, with some exceptions, apply as bounds for $G_{\text{ineq}}(k)$.

As Davenport notes, the proof in Chapter 20 only applies to a sequence of large $P$, where the sequence depends on the rational approximation properties of the ratios of the coefficients. In many applications of the

Hardy–Littlewood method, one obtains an asymptotic formula for the number of integral solutions for all positive $P$ with not much more effort than is required to establish solubility. For example, for indefinite diagonal forms with coefficients nonzero and not all in rational ratio, and for positive $P$, and $s$ sufficiently large in terms of $k$, we would expect that the number $N(P)$ of integral solutions $\mathbf{x}$ of (4) with $|x_i| \le P$ for $1 \le i \le s$ satisfies

$$N(P) = C(s,k,\lambda_1,\ldots,\lambda_s)\varepsilon P^{s-k} + o\left(P^{s-k}\right), \tag{6}$$

where $C(s,k,\lambda_1,\ldots,\lambda_s)$ is a positive constant depending only on $s,k$ and the coefficients $\lambda_i$. However, the proof of Davenport and Heilbronn (with some minor technical modifications) allows one to give asymptotic formulae for diagonal Diophantine inequalities for essentially only an infinite sequence of large $P$. In their paper, Bentkus and Götze [3] establish the appropriate analogue of (6) for general positive definite quadratic forms with coefficients not all in rational ratio, for all positive $P$; although their proofs are not phrased in the language of the Davenport–Heilbronn method, the ideas are similar. By adapting their work, Freeman [33, 35] was able to prove the existence of an asymptotic formula such as (6) for indefinite diagonal forms of degree $k$ for all positive $P$. Wooley [101] has considerably simplified and improved this work, using clever ideas to reduce the number of variables needed to guarantee the existence of asymptotic formulae.

In particular, for the existence of asymptotic formulae for large $k$, one can establish results similar to (5); if we define $G_{\text{asymp}}(k)$ analogously to $G_{\text{ineq}}(k)$, one has

$$G_{\text{asymp}}(k) \le k^2 \left(\log k + \log\log k + O(1)\right).$$

Finally, we note that Eskin, Margulis and Mozes [31], using techniques different from the Davenport–Heilbronn method, in fact earlier proved the existence of asymptotic formulae of the expected kind for the case of general indefinite quadratic forms in at least four variables with coefficients not all in rational ratio, and signature not equal to $(2,2)$.

D. E. Freeman
*Carleton University*

# Editorial preface

Like many mathematicians I first came into contact with number theory through Davenport's book *The Higher Arithmetic* [23]. It was difficult not to be struck by his command of the subject and wonderful expository style. This basic textbook is now into its seventh edition, whilst at a more advanced level, a third edition of Davenport's *Multiplicative Number Theory* [24] has recently appeared. It is fair to say therefore that Davenport still holds considerable appeal to mathematicians worldwide. On discovering that Davenport had also produced a rather less well-known set of lecture notes treating an area of substantial current interest, I was immediately compelled to try and get it back into print. In doing so, I have tried to preserve in its original format as much of the material as possible, and have merely removed errors that I encountered along the way.

As the title indicates, this book is concerned with the use of analytic methods in the study of integer solutions to certain polynomial equations and inequalities. It is based on lectures that Davenport gave at the University of Michigan in the early 1960s. This analytic method is usually referred to as the 'Hardy–Littlewood circle method', and its power is readily demonstrated by the diverse range of number theoretic problems that can be tackled by it. The first half of the book is taken up with a discussion of the method in its most classical setting: Waring's problem and the representation of integers by diagonal forms. In Chapters 11–19, Davenport builds upon these foundations by showing how the method can sometimes be adapted to handle integer solutions of general systems of homogeneous polynomial equations. Finally, in Chapter 20 Davenport presents an account of work carried out by himself and Heilbronn in the setting of Diophantine inequalities. Even more so than with his other books, these lecture notes reflect Davenport's extensive influence in the subject area and his deep knowledge pertaining to it.

This edition of Davenport's lecture notes has been considerably enriched by the provision of a foreword, the main purpose of which is to place a modern perspective on the state of knowledge described in the lecture notes. I am extremely grateful to Professor Freeman, Professor Heath-Brown and Professor Vaughan for lending their authority to this project. I also wish to thank Lillian Pierce and Luke Woodward for all of their hard work in helping me transcribe Davenport's original lecture notes into LaTeX. Finally it is a pleasure to express my gratitude both to James Davenport at Bath University and to David Tranah at Cambridge University Press for sharing my enthusiasm in bringing these lecture notes to the attention of a wider mathematical audience.

T. D. Browning
*Mathematical Institute*
*Oxford University*
*24–29 St. Giles'*
*Oxford*
*OX1 3LB*
browning@maths.ox.ac.uk

# 1

# Introduction

The analytic method of Hardy and Littlewood (sometimes called the 'circle method') was developed for the treatment of *additive problems* in the theory of numbers. These are problems which concern the representation of a large number as a sum of numbers of some specified type. The number of summands may be either fixed or unrestricted; in the latter case we speak of *partition problems.* The most famous additive problem is Waring's problem, where the specified numbers are the $k$th powers, so that the problem is that of representing a large number $N$ as

$$N = x_1^k + x_2^k + \cdots + x_s^k, \tag{1.1}$$

where $s$ and $k$ are given and $x_1, \ldots, x_s$ are positive integers. Almost equally famous is Goldbach's ternary problem, where the specified numbers are the primes, and the problem is that of representing a large number $N$ as

$$N = p_1 + p_2 + p_3.$$

The great achievements of Hardy and Littlewood were followed later by further remarkable progress made by Vinogradov, and it is not without justice that our Russian colleagues now speak of the 'Hardy–Littlewood–Vinogradov method'.

It may be of interest to recall that the genesis of the Hardy–Littlewood method is to be found in a paper of Hardy and Ramanujan [69] in 1917 on the asymptotic behaviour of $p(n)$, the total number of partitions of $n$. The function $p(n)$ increases like $e^{A\sqrt{n}}$, where $A$ is a certain positive constant; and Hardy and Ramanujan obtained for it an asymptotic series, which, if one stops at the smallest term, gives $p(n)$ with an error $O(n^{-1/4})$. The underlying explanation for this high degree of accuracy, which Hardy describes as 'uncanny', was given by Rademacher [68] in 1937: there is a convergent series which represents $p(n)$ exactly, and this is initially almost the same as the asymptotic series. There is one other group of problems in which the Hardy–Littlewood method leads to exact formulae; these are problems concerning the representation of a number as the sum of a given number of squares. It seems unlikely that there are any such formulae for higher powers.

Waring's problem is concerned with the particular Diophantine equation (1.1). There is no difficulty of principle in extending the Hardy–Littlewood method to deal with more general equations of additive type[1], say

$$N = f(x_1) + f(x_2) + \cdots + f(x_s),$$

where $f(x)$ is a polynomial taking integer values; in particular to the equation

$$N = a_1 x_1^k + a_2 x_2^k + \cdots + a_s x_s^k. \tag{1.2}$$

It is only in recent years, however, that much progress has been made in adapting the method to Diophantine equations of a general (that is, non-additive) character. An account of these developments will be given later in these lectures, but we shall be concerned at first mainly with Waring's problem and with additive equations of the type (1.2). All work on general Diophantine equations depends heavily on either the methods or the results of the work on additive equations.

Finally, we shall touch on the subject of Diophantine inequalities. Here, too, some results of a general character are now known, but they are less complete and less precise than those for equations.

[1] See the monograph [63].

# 2

# Waring's problem: history

In his *Meditationes algebraicae* (1770), Edward Waring made the statement that every number is expressible as a sum of 4 squares, or 9 cubes, or 19 biquadrates, 'and so on'. By the last phrase, it is presumed that he meant to assert that for every $k \geq 2$ there is some $s$ such that every positive integer $N$ is representable as

$$N = x_1^k + x_2^k + \cdots + x_s^k, \tag{2.1}$$

for $x_i \geq 0$. This assertion was first proved by Hilbert in 1909. Hilbert's proof was a very great achievement, though some of the credit should go also to Hurwitz, whose work provided the starting point. Hurwitz had already proved that if the assertion is true for any exponent $k$, then it is true for $2k$. I shall not discuss Hilbert's method of proof here; for this one may consult papers by Stridsberg [79], Schmidt [72] or Rieger [71]. It is usual to denote the least value of $s$, such that every $N$ is representable, by $g(k)$. The exact value of $g(k)$ is now known for all values of $k$.

The work of Hardy and Littlewood appeared in several papers of the series 'On Partitio Numerorum' (P.N.), the other papers of the series being concerned mainly with Goldbach's ternary problem. In P.N. I [37] they obtained an asymptotic formula for $r(N)$, the number of representations of $N$ in the form (2.1) with $x_i \geq 1$, valid provided $s \geq s_0(k)$, a certain explicit function of $k$. The asymptotic formula was of the following form:

$$r(N) = C_{k,s} N^{s/k-1} \mathfrak{S}(N) + O(N^{s/k-1-\delta}), \tag{2.2}$$

where $\delta > 0$ and

$$C_{k,s} = \frac{\Gamma(1+1/k)^s}{\Gamma(s/k)} > 0.$$

In the above formula, $\mathfrak{S}(N)$ is an infinite series of a purely arithmetical nature, which Hardy and Littlewood called the *singular series*. They proved further that

$$\mathfrak{S}(N) \geq \gamma > 0, \tag{2.3}$$

for some $\gamma$ independent of $N$, provided that $s \geq s_1(k)$. However they did not at that stage give any explicit value for $s_1(k)$. Thus the formula implies that

$$r(N) \sim C_{k,s} N^{s/k-1} \mathfrak{S}(N) \tag{2.4}$$

as $N \to \infty$, provided $s \geq \max(s_0(k), s_1(k))$, and thereby provided an independent proof of Hilbert's theorem.

Hardy and Littlewood introduced the notation $G(k)$ for the least value of $s$ such that every sufficient large $N$ is representable in the form (2.1); this function is really of more significance than $g(k)$, since the latter is affected by the difficulty of representing one or two particular numbers $N$. In P.N. II [38] and P.N. IV [39], Hardy and Littlewood proved that the asymptotic formula and the lower bound for $\mathfrak{S}(N)$ both hold for $s \geq (k-2)2^{k-1} + 5$, which implies that

$$G(k) \leq (k-2)2^{k-1} + 5.$$

In P.N. VI [40] they found a better upper bound for $G(k)$, though not for the validity of the asymptotic formula, and in particular they proved that $G(4) \leq 19$. The last paper of the series, P.N. VIII [41], was entirely concerned with the singular series and with the congruence problem to which it gives rise.

Hardy and Littlewood took as their starting point the generating function for $r(N)$, that is, the power series

$$\sum_{N=0}^{\infty} r(N) z^N = \left( \sum_{n=0}^{\infty} z^{n^k} \right)^s.$$

They expressed $r(N)$ in terms of this function by means of Cauchy's formula for the coefficients of a power series, using a contour integral taken along the circle $|z| = \rho$, where $\rho$ is slightly less than 1. A helpful technical simplification was introduced by Vinogradov in 1928; this consists of replacing the power series by a finite exponential sum, and the effect is to eliminate a number of unimportant complications that occurred in the original presentation of Hardy and Littlewood.

Write $e(t) = e^{2\pi it}$. We define $T(\alpha)$, for a real variable $\alpha$, by

$$T(\alpha) = \sum_{x=1}^{P} e(\alpha x^k), \tag{2.5}$$

where $P$ is a positive integer. Then

$$(T(\alpha))^s = \sum_{m} r'(m)e(m\alpha), \tag{2.6}$$

where $r'(m)$ denotes the number of representations of $m$ as

$$x_1^k + \cdots + x_s^k, \quad (1 \leq x_i \leq P).$$

If $P \geq [N^{1/k}]$, where $[\lambda]$ denotes the integer part of any real number $\lambda$, then $r'(N)$ is the total number of representations of $N$ in the form (2.1) with $x_i \geq 1$. Consequently $r'(N) = r(N)$. If we multiply both sides of (2.6) by $e(-N\alpha)$ and integrate over the unit interval $[0, 1]$ (or over any interval of length 1), we get

$$r(N) = \int_0^1 (T(\alpha))^s e(-N\alpha)d\alpha. \tag{2.7}$$

This is the starting point of our work on Waring's problem. It corresponds to the contour integral for $r(N)$ used by Hardy and Littlewood, with $z$ replaced by $e^{2\pi i\alpha}$.

Our first aim will be to establish the validity of the asymptotic formula (2.2) for $r(N)$ as $N \to \infty$, subject to the condition $s \geq 2^k + 1$. It is possible to do this in a comparatively simple manner by using an inequality found by Hua in 1938 (Lemma 3.2 below). It may be of interest to observe that no improvement on the condition $s \geq 2^k + 1$ has yet been made for small values of $k$, as far as the asymptotic formula itself is concerned. For large $k$ it has been shown by Vinogradov that a condition of the type $s > Ck^2 \log k$ is sufficient.

If we prove that the asymptotic formula holds for a particular value of $s$, say $s = s_1$, it will follow that every large number is representable as a sum of $s_1$ $k$th powers, whence $G(k) \leq s_1$. But to prove this it is not essential to prove the asymptotic formula for the total number of representations; it would be enough to prove it for some special type of representation as a sum of $s_1$ $k$th powers. This makes it possible to get better estimates for $G(k)$ than one can get for the validity of the asymptotic formula. In 1934 Vinogradov proved that $G(k) < Ck \log k$ for large $k$, and we shall give a proof in Chapter 9. The best known results for small $k$ were found by Davenport in 1939–41 [19].

A new 'elementary' proof of Hilbert's theorem was given by Linnik in 1943 [58], and was selected by Khintchine as one of his 'three pearls' [53]. The underlying ideas of this proof were undoubtedly suggested by certain features of the Hardy–Littlewood method, and in particular by Hua's inequality.

# 3

# Weyl's inequality and Hua's inequality

The most important single tool for the investigation of Waring's problem, and indeed many other problems in the analytic theory of numbers, is Weyl's inequality. This was given, in a less explicit form, in Weyl's great memoir [96] of 1916 on the uniform distribution of sequences of numbers to the modulus 1. The explicit form for a polynomial, in terms of a rational approximation to the highest coefficient, was given by Hardy and Littlewood in P.N. I [37].

**Lemma 3.1. (Weyl's Inequality)** *Let $f(x)$ be a real polynomial of degree $k$ with highest coefficient $\alpha$:*

$$f(x) = \alpha x^k + \alpha_1 x^{k-1} + \cdots + \alpha_k.$$

*Suppose that $\alpha$ has a rational approximation $a/q$ satisfying*

$$(a,q) = 1, \quad q > 0, \quad \left|\alpha - \frac{a}{q}\right| \le \frac{1}{q^2}.$$

*Then, for any $\varepsilon > 0$,*

$$\left|\sum_{x=1}^{P} e(f(x))\right| \ll P^{1+\varepsilon}\left(P^{-\frac{1}{K}} + q^{-\frac{1}{K}} + \left(\frac{P^k}{q}\right)^{-\frac{1}{K}}\right),$$

*where $K = 2^{k-1}$ and the implied constant*[1] *depends only on $k$ and $\varepsilon$.*

**Note.** The inequality gives *some* improvement on the trivial upper bound $P$ provided that $P^\delta \le q \le P^{k-\delta}$ for some fixed $\delta > 0$. If $P \le q \le P^{k-1}$, we get the estimate $P^{1-1/K+\varepsilon}$, and it is under these

[1] We use the Vinogradov symbol $\ll$ to indicate an inequality with an unspecified 'constant' factor. In the present instance, the factor which arises is in reality independent of $k$, but we do not need to know this.

conditions that Weyl's inequality is most commonly applied. It is obviously impossible to extract any better estimate than this from it. Note that Weyl's inequality fails to give any useful information if $q$ is small, and this is natural because if $f(x) = \alpha x^k$ and $\alpha$ is very near to a rational number with small denominator, the sum is genuinely of a size which approaches $P$.

*Proof.* The basic operation in the proof is that of squaring the absolute value of an exponential sum, and thereby relating the sum to an average of similar sums with polynomials of degree one lower. Let

$$S_k(f) = \sum_{x=P_1+1}^{P_2} e(f(x)),$$

where $0 \le P_2 - P_1 \le P$, and where the suffix $k$ serves to indicate the degree of $f(x)$. Then

$$\begin{aligned} |S_k(f)|^2 &= \sum_{x_1}\sum_{x_2} e(f(x_2) - f(x_1)) \\ &= P_2 - P_1 + 2\Re \sum_{\substack{x_1, x_2 \\ x_2 > x_1}} e(f(x_2) - f(x_1)). \end{aligned}$$

Put $x_2 = x_1 + y$. Then $1 \le y < P_2 - P_1$, and

$$f(x_2) - f(x_1) = f(x_1 + y) - f(x_1) = \Delta_y f(x_1),$$

with an obvious notation. Hence

$$|S_k(f)|^2 = P_2 - P_1 + 2\Re \sum_{y=1}^{P} \sum_{x} e\left(\Delta_y f(x)\right),$$

where the summation in $x$ is over an interval depending on $y$ but contained in $P_1 < x \le P_2$. This interval may, for some values of $y$, be empty.

In particular,

$$|S_k(f)|^2 \le P + 2\sum_{y=1}^{P} |S_{k-1}(\Delta_y f)|,$$

where the interval for $S_{k-1}$ is of the nature just described. By repeating the argument we get

$$|S_{k-1}(\Delta_y f)|^2 \leq P + 2\sum_{z=1}^{P} |S_{k-2}(\Delta_{y,z} f)|,$$

where the interval of summation in $S_{k-2}$ depends on both $y$ and $z$ but is contained in $P_1 < x \leq P_2$. The use of Cauchy's inequality enables us to substitute for $S_{k-1}$ from the second inequality into the first:

$$\begin{aligned} |S_k(f)|^4 &\ll P^2 + P\sum_{y=1}^{P} |S_{k-1}(\Delta_y f)|^2 \\ &\ll P^3 + P\sum_{y=1}^{P}\sum_{z=1}^{P} |S_{k-2}(\Delta_{y,z} f)|. \end{aligned}$$

The process can be continued, and the general inequality established in this way is

$$|S_k(f)|^{2^\nu} \ll P^{2^\nu - 1} + P^{2^\nu - \nu - 1} \sum_{y_1=1}^{P} \cdots \sum_{y_\nu=1}^{P} |S_{k-\nu}(\Delta_{y_1,\ldots,y_\nu} f)|. \quad (3.1)$$

This is readily proved by induction on $\nu$, using again Cauchy's inequality together with the basic operation described above which expresses $|S_{k-\nu}|^2$ in terms of $S_{k-\nu-1}$. It is to be understood that the range of summation for $x$ in $S_{k-\nu}$ in (3.1) is an interval depending on $y_1, \ldots, y_\nu$, but contained in $P_1 < x \leq P_2$.

At this point we interpolate a remark which will be useful in the proof of Lemma 3.2. This is that if, at the last stage of the proof of (3.1), we apply the basic operation in its original form, we get

$$|S_k(f)|^{2^\nu} \ll P^{2^\nu - 1} + P^{2^\nu - \nu - 1} \sum_{y_1=1}^{P} \cdots \sum_{y_\nu=1}^{P} \Re S_{k-\nu}(\Delta_{y_1,\ldots,y_\nu} f). \quad (3.2)$$

Here again, the range for $x$ in $S_{k-\nu}$ depends on $y_1, \ldots, y_\nu$ and may sometimes be empty.

Returning to (3.1), we take $\nu = k-1$ and in the original $S_k$ we take $P_1 = 0$, $P_2 = P$. We observe that

$$\Delta_{y_1,\ldots,y_{k-1}} f(x) = k!\alpha y_1 \cdots y_{k-1} x + \beta,$$

say, where $\beta$ is a collection of terms independent of $x$. Hence

$$|S_1(\Delta_{y_1,\ldots,y_{k-1}} f)| = \left|\sum_x e(k!\alpha y_1 \cdots y_{k-1} x)\right|.$$

The sum on the right, taken over any interval of $x$ of length at most $P$, is of the form

$$\left|\sum_{x=x_1}^{x_2-1} e(\lambda x)\right| \le \frac{2}{|1-e(\lambda)|} = \frac{1}{|\sin \pi\lambda|} \ll \frac{1}{\|\lambda\|},$$

where $\|\lambda\|$ denotes the distance of $\lambda$ from the nearest integer. This fails if $\lambda$ is an integer, and indeed gives a poor result if $\lambda$ is very near to an integer, but we can supplement it by the obvious upper bound $P$. Hence (3.1) gives

$$|S_k(f)|^K \ll P^{K-1} + P^{K-k} \sum_{y_1=1}^{P} \cdots \sum_{y_{k-1}=1}^{P} \min(P, \|k!\alpha y_1 \cdots y_{k-1}\|^{-1}).$$

We now appeal to a result in elementary number theory, which enables us to collect together all the terms in the sum for which $k! y_1 \cdots y_{k-1}$ has a given value, say $m$. The number of such terms is $\ll m^{\varepsilon}$. To prove this, it suffices to show that

$$d(m) \ll m^{\varepsilon}, \tag{3.3}$$

for any integer $m$, where $d(m) = \sum_{d|m} 1$ is the usual divisor function. Indeed there are at most $d(m)$ possibilities for each of $y_1, \ldots, y_{k-1}$. To establish (3.3) we suppose that $m = p_1^{\lambda_1} p_2^{\lambda_2} \cdots$, and note that

$$\frac{d(m)}{m^{\varepsilon}} = \prod_i \frac{\lambda_i + 1}{p_i^{\varepsilon\lambda_i}} \le \prod_{p_i \le 2^{1/\varepsilon}} \frac{\lambda_i + 1}{2^{\varepsilon\lambda_i}} \le C(\varepsilon),$$

since $2^{-\varepsilon\lambda}(\lambda+1)$ is bounded above for $\lambda > 0$.

Collecting terms as mentioned above, we get

$$|S_k(f)|^K \ll P^{K-1} + P^{K-k+\varepsilon} \sum_{m=1}^{k!P^{k-1}} \min(P, \|\alpha m\|^{-1}).$$

It remains to estimate the last sum in terms of the rational approximation $a/q$ to $\alpha$ which was mentioned in the enunciation. We divide the sum over $m$ into blocks of $q$ consecutive terms (with perhaps one incomplete block), the number of such blocks being

$$\ll \frac{P^{k-1}}{q} + 1.$$

Consider the sum over any one block, which will be of the form

$$\sum_{m=0}^{q-1} \min(P, \|\alpha(m_1 + m)\|^{-1}),$$

where $m_1$ is the first number in the block. We have

$$\alpha(m_1 + m) = \alpha m_1 + \frac{am}{q} + O\left(\frac{1}{q}\right),$$

since $|\alpha - a/q| \leq q^{-2}$ and $0 \leq m < q$. As $m$ goes from 0 to $q-1$, the number $am$ runs through the complete set of residues (mod $q$). Putting $am \equiv r \pmod q$, the sum is

$$\sum_{r=0}^{q-1} \min\left(P, \frac{1}{\|(r+b)/q + O(1/q)\|}\right),$$

where we have taken $b$ to be the integer nearest to $q\alpha m_1$. There are $O(1)$ values of $r$ in the sum for which the second expression in the minimum is useless, namely those for which the absolutely least residue of $r+b$ (mod $q$) is small. For these, we must take $P$. For the other values of $r$, if $s$ denotes the absolutely least residue of $r+b$ (mod $q$) we have

$$\left\|\frac{r+b}{q} + O\left(\frac{1}{q}\right)\right\| \gg \frac{s}{q}.$$

Hence the above sum is

$$\ll P + \sum_{s=1}^{q/2} \frac{q}{s} \ll P + q\log q.$$

Allowing for the number of blocks, we obtain

$$|S_k(f)|^K \ll P^{K-1} + P^{K-k+\varepsilon}\left(\frac{P^{k-1}}{q} + 1\right)(P + q\log q).$$

We can absorb the factor $\log q$ in $P^\varepsilon$, since we can suppose $q \leq P^k$, as otherwise the result of the lemma is trivial. Thus the right-hand side is

$$\ll P^{K+\varepsilon}\left(P^{-1} + q^{-1} + P^{-k}q\right),$$

giving the result. □

**Note.** If $k$ is large, then Vinogradov has given a much better estimate, in which (roughly speaking) $2^{k-1}$ is replaced by $4k^2\log k$ [49, Chapter 6].

**Corollary (to Lemma 3.1).** *Let*

$$S_{a,q} = \sum_{z=1}^{q} e(az^k/q),$$

*where $a$, $q$ are relatively prime integers and $q > 0$. Then*

$$S_{a,q} \ll q^{1-1/K+\varepsilon}.$$

This is a special case of Lemma 3.1 with $\alpha = a/q$ and $P = q$. We shall later (Lemma 6.4) prove the more precise estimate $q^{1-1/k}$ instead of $q^{1-1/K+\varepsilon}$, but the above suffices for the time being.

**Lemma 3.2. (Hua's Inequality [48])** *If*

$$T(\alpha) = \sum_{x=1}^{P} e(\alpha x^k),$$

*then*

$$\int_0^1 |T(\alpha)|^{2^k} \, d\alpha \ll P^{2^k-k+\varepsilon}$$

*for any fixed $\varepsilon > 0$.*

*Proof.* Write

$$I_\nu = \int_0^1 |T(\alpha)|^{2^\nu} \, d\alpha.$$

We prove, by induction on $\nu$, that

$$I_v \ll P^{2^\nu-\nu+\varepsilon}, \quad \text{for } \nu = 1, \dots, k, \tag{3.4}$$

the case $\nu = k$ being the result asserted in the lemma.

For $\nu = 1$, the estimate is immediate. We have

$$I_1 = \int_0^1 \sum_{x_1} e(\alpha x_1^k) \sum_{x_2} e(-\alpha x_2^k) \, d\alpha = P,$$

since the integral over $\alpha$ is 1 if $x_1 = x_2$ and 0 otherwise.

Now suppose (3.4) holds for a particular integer $\nu \le k-1$; we have to deduce the corresponding result when $\nu$ is replaced by $\nu+1$. We recall the inequality (3.2) of the preceding proof; with $T(\alpha)$ in place of $S_k(f)$ it states that

$$|T(\alpha)|^{2^\nu} \ll P^{2^\nu-1} + P^{2^\nu-\nu-1} \Re \sum_{y_1=1}^{P} \cdots \sum_{y_\nu=1}^{P} S_{k-\nu},$$

where

$$S_{k-\nu} = \sum_x e(\alpha \Delta_{y_1,\dots,y_\nu}(x^k)).$$

Note that the range of summation for $x$ depends on the values of $y_1, \ldots, y_\nu$, but is contained in $[1, P]$.

Multiply both sides of the inequality by $|T(\alpha)|^{2^\nu}$ and integrate from 0 to 1. We get

$$I_{\nu+1} \ll P^{2^\nu-1} I_\nu + P^{2^\nu-\nu-1} \sum_{y_1,\ldots,y_\nu} \Re \int_0^1 S_{k-v} |T|^{2^\nu} \, d\alpha.$$

The last integral is

$$\int_0^1 \sum_x e\left(\alpha \Delta_{y_1,\ldots,y_\nu}(x^k)\right) \sum_{\substack{u_1,\ldots,u_{2^\nu-1} \\ v_1,\ldots,v_{2^\nu-1}}} e(\alpha u_1^k + \cdots) e(-\alpha v_1^k - \cdots) \, d\alpha,$$

where the $u_i$ and $v_i$ go from 1 to $P$. This integral equals the number of solutions of

$$\Delta_{y_1,\ldots,y_\nu}(x^k) + u_1^k + \cdots - v_1^k - \cdots = 0. \tag{3.5}$$

Summation over $y_1, \ldots, y_\nu$ gives the number of solutions in all the variables. Hence

$$I_{\nu+1} \ll P^{2^\nu-1} I_\nu + P^{2^\nu-\nu-1} N, \tag{3.6}$$

where $N$ denotes the number of solutions of (3.5) in all the variables, these being now any integers in $[1, P]$.

It is important now to observe that since $y_1, \ldots, y_\nu$ and $x$ are positive, we have

$$\Delta_{y_1,\ldots,y_\nu}(x^k) > 0.$$

Also, this number is divisible by each of $y_1, \ldots, y_\nu$. Thus, if we give $u_1, \ldots, u_{2^\nu-1}$ and $v_1, \ldots, v_{2^\nu-1}$ any values, the number of possibilities for each of $y_1, \ldots, y_\nu$ is $\ll P^\varepsilon$ by (3.3). Then there is at most one possibility for $x$, since $\Delta_{y_1,\ldots,y_\nu}(x^k)$ is a strictly increasing function of $x$ (note that $\nu \leq k-1$). The number of possibilities for the $u_i$ and $v_i$ is $\ll P^{2^\nu}$, whence it follows that

$$N \ll P^{2^\nu+\nu\varepsilon}.$$

Substituting in (3.6) and using the inductive hypothesis, we obtain

$$I_{\nu+1} \ll P^{2^\nu-1} P^{2^\nu-\nu+\varepsilon} + P^{2^\nu-\nu-1} P^{2^\nu+\nu\varepsilon} \ll P^{2^{\nu+1}-(\nu+1)+\nu\varepsilon}.$$

This is (3.4) with $\nu+1$ for $\nu$, except for the change in $\varepsilon$ which is of no significance. □

**Note.** It is of interest to examine the information given by Lemma 3.2 when $k = 3$. Let $\lambda(m)$ denote the lower bound of the exponents $\lambda$ for which it is true that

$$\int_0^1 |T(\alpha)|^{2m} \, d\alpha \ll P^{\lambda}.$$

It follows from Cauchy's inequality that

$$\lambda\left(\frac{m_1 + m_2}{2}\right) \leq \frac{1}{2}(\lambda(m_1) + \lambda(m_2)),$$

so that the graph of $\lambda(m)$ as a function of $m$ is convex. Lemma 3.2 tells us that

$$\lambda(1) \leq 1, \quad \lambda(2) \leq 2, \quad \lambda(4) \leq 5,$$

and it can be proved that actually there is equality in all these. Thus the graph lies on or below the two line segments joining $(1,1)$, $(2,2)$, $(4,5)$. It seems likely that the graph is strictly below the segment for $2 < m < 4$, but this has never been proved. If it could be proved, one could establish the asymptotic formula for eight cubes instead of for nine cubes $(9 = 2^k + 1)$. It would be enough to prove, for example, that

$$\int_0^1 |T(\alpha)|^6 \, d\alpha \ll P^{7/2-\delta}$$

for some positive $\delta$. This is equivalent to the assertion that the total number of solutions of

$$x_1^3 + x_2^3 + x_3^3 = y_1^3 + y_2^3 + y_3^3,$$

with all the variables between 0 and $P$, is $\ll P^{7/2-\delta}$.

# 4

# Waring's problem: the asymptotic formula

We return to the starting point for our work on Waring's problem, namely (2.7) of Chapter 2:

$$r(N) = \int_0^1 (T(\alpha))^s e(-N\alpha) d\alpha, \tag{4.1}$$

where $T(\alpha)$ is the exponential sum (2.5) from 1 to $P$ and $P \geq [N^{1/k}]$. There is no point in taking $P$ larger than necessary, so we take $P = [N^{1/k}]$. The main term in the asymptotic formula will prove to be of order $N^{s/k-1}$, or $P^{s-k}$, as indeed it must be if any simple asymptotic formula is valid, for this is the only power of $P$ which is consistent with the fact that there are $P^s$ choices for $x_1, \ldots, x_s$ and the sums $x_1^k + \cdots + x_s^k$ represent numbers of order at most $P^k$.

Thus we can neglect any set of values of $\alpha$ in the integral (4.1) which can be shown to contribute to the integral an amount which is of strictly lower order than $P^{s-k}$. We are supposing $s \geq 2^k + 1$, and if we regard the absolute value of the integrand as

$$|T(\alpha)|^{s-2^k} |T(\alpha)|^{2^k},$$

it will follow from Lemma 3.2 that we can neglect any set of $\alpha$ for which $|T(\alpha)| \ll P^{1-\delta}$ for some fixed $\delta > 0$. To obtain such a set of $\alpha$, we shall use Lemma 3.1.

The general plan in work on Waring's problem and similar problems is to divide the values of $\alpha$ into two sets: the *major arcs*, which contribute to the main term in the asymptotic formula, and the *minor arcs*, the contribution of which is estimated on lines such as those described above, and goes into the error term. The precise line of demarcation between the two sets depends very much on what particular auxiliary results are available, and may to some extent be a matter of personal choice. Generally speaking there are powerful (though somewhat complicated) methods available for the treatment of the major arcs, and the crux of the problem lies with the minor arcs. Having found, in any particular problem, a method which copes successfully with the minor arcs, one usually finds it convenient to enlarge them as far as the method in question will permit, in order to reduce the amount of work needed for the major arcs (even though this work might be relatively straightforward). In the present treatment we can take the major arcs to be few in number and short in length, compared with what is often the case in other work on the subject.

Around every rational number $a/q$ (in its lowest terms) we put an interval

$$\mathfrak{M}_{a,q}: \quad |\alpha - a/q| < P^{-k+\delta}, \tag{4.2}$$

and we do this for

$$1 \leq q \leq P^{\delta}, \quad 1 \leq a \leq q, \quad (a,q) = 1. \tag{4.3}$$

These intervals do not overlap, since the distance between their centres is at least $P^{-2\delta}$ and this is much greater than their length. Moreover these intervals are contained in $0 \leq \alpha \leq 1$ except for the right-hand half of the interval round $1/1$, and for convenience we imagine this interval translated an amount 1 to the left, so that it comes to the right of $\alpha = 0$. The intervals $\mathfrak{M}_{a,q}$ are the major arcs, and their complement relative to $[0,1]$ constitutes the minor arcs, the totality of which we shall denote by $\mathfrak{m}$. In these definitions, $\delta$ is some fixed small positive number. It may be remarked that in many applications of the Hardy–Littlewood method, the length of $\mathfrak{M}_{a,q}$ in (4.2) would incorporate a factor $q^{-1}$ as well as a negative power of $P$, but here this factor is not needed and it is a slight simplification to omit it.

**Lemma 4.1.** *If* $s \geq 2^k + 1$, *we have*

$$\int_{\mathfrak{m}} |T(\alpha)|^s d\alpha \ll P^{s-k-\delta'}$$

*where* $\delta'$ *is a positive number depending on* $\delta$.

*Proof.* By a classical result of Dirichlet on Diophantine approximations, every $\alpha$ has a rational approximation $a/q$ satisfying

$$1 \leq q \leq P^{k-\delta}, \quad |\alpha - a/q| < q^{-1}P^{-k+\delta}. \tag{4.4}$$

Moreover, we have $1 \leq a \leq q$ whenever $0 < \alpha < 1$. Since the last inequality in (4.4) is stronger than that in (4.2), we should have $\alpha$ in $\mathfrak{M}_{a,q}$ if $q \le P^{\delta}$. Hence if $\alpha$ is in $\mathfrak{m}$, we must have

$$q > P^{\delta}.$$

Since $|\alpha - a/q| < q^{-2}$, we can apply Lemma 3.1 to the exponential sum $T(\alpha)$, and since $P^k/q \ge P^{\delta}$ we get

$$|T(\alpha)| \ll P^{1+\varepsilon-\delta/K},$$

where $K = 2^{k-1}$. Combining this with Lemma 3.2, in the manner indicated earlier, we infer that

$$\int_{\mathfrak{m}} |T(\alpha)|^s d\alpha \ll P^{(s-2^k)(1+\varepsilon-\delta/K)} \int_0^1 |T(\alpha)|^{2^k} d\alpha$$
$$\ll P^{s-k-\delta'}$$

for some $\delta' > 0$ depending on $\delta$. This proves Lemma 4.1. □

It may be noted that instead of appealing to Dirichlet's theorem we could use a simple property of continued fractions: if we take $a/q$ to be the last convergent to $\alpha$ for which $q \le P^{k-\delta}$, we again get (4.4).

We now turn our attention to the major arcs $\mathfrak{M}_{a,q}$. Here $\alpha$ is very near to $a/q$, with $q$ relatively small. If the exponent of $P$ in (4.2) had been $-k-\delta$ instead of $-k+\delta$, then $T(\alpha)$ would be practically constant on $\mathfrak{M}_{a,q}$, for we should have

$$\left|\alpha x^k - \frac{a}{q}x^k\right| < P^{-k-\delta}P^k = P^{-\delta}.$$

This, of course, is not the case, but nevertheless, the arc $\mathfrak{M}_{a,q}$ is so short that $T(\alpha)$ behaves relatively smoothly in that interval. Just how it varies is seen in the following lemma.

**Lemma 4.2.** *For $\alpha$ in $\mathfrak{M}_{a,q}$, putting $\alpha = \beta + a/q$, we have*

$$T(\alpha) = q^{-1}S_{a,q}I(\beta) + O(P^{2\delta}), \tag{4.5}$$

*where*

$$S_{a,q} = \sum_{z=1}^{q} e(az^k/q), \tag{4.6}$$

$$I(\beta) = \int_0^P e(\beta\xi^k)d\xi. \tag{4.7}$$

*Proof.* We collect together those values of $x$ in the sum defining $T(\alpha)$ which are in the same residue class (mod $q$), as is natural because $\alpha x^k$ is not far from being periodic in $x$ with period $q$. This is most conveniently done by putting $x = qy + z$ where $1 \le z \le q$; here $y$ runs through an interval, depending on $z$, corresponding to the interval $0 < x \le P$. We get

$$T(\alpha) = \sum_{z=1}^{q} e(az^k/q) \sum_{y} e(\beta(qy+z)^k).$$

Now we endeavour to replace the discrete variable $y$ by a continuous variable $\eta$, and replace the summation over $y$ by an integration over $\eta$. If this can be done, we can then make a change of variable from $\eta$ to $\xi$, where $\xi = q\eta + z$; the interval for $\xi$ will be the original interval $0 \le \xi \le P$, and we shall have replaced the summation over $y$ by

$$q^{-1} \int_0^P e(\beta \xi^k) d\xi = q^{-1} I(\beta),$$

the factor $q^{-1}$ coming from $d\eta/d\xi$. Thus we shall get precisely the main term in (4.5).

We have to estimate the difference between the sum over $y$ and the corresponding integral over $\eta$. For the present purpose a very crude argument is good enough. If $f(y)$ is any differentiable function, we have

$$|f(\eta) - f(y)| \le \frac{1}{2} \max |f'(\eta)| \quad \text{for} \quad |\eta - y| \le \frac{1}{2}.$$

Hence, on dividing any interval $A < \eta < B$ into intervals of length 1 together with two possible broken intervals, we obtain

$$\left| \int_A^B f(\eta) d\eta - \sum_{A<y<B} f(y) \right| \ll (B-A) \max |f'(\eta)| + \max |f(\eta)|.$$

In our case, $f(\eta) = e(\beta(q\eta + z)^k)$, whence

$$\max |f'(\eta)| \ll q|\beta| P^{k-1}, \quad \max |f(\eta)| = 1.$$

Also $B - A \ll P/q$. Hence the error in replacing the sum over $y$ by the integral over $\eta$ is

$$\ll Pq^{-1} q |\beta| P^{k-1} + 1 \ll P^{\delta},$$

since $|\beta| < P^{-k+\delta}$ by (4.2). Multiplying by $q$, which is $\le P^{\delta}$, to allow for the outside summation over $z$, we obtain the error term in (4.5). □

Later, in Lemma 9.1, we shall meet a more effective method for replacing a sum by the corresponding integral.

**Lemma 4.3.** *If* $\mathfrak{M}$ *denotes the totality of the major arcs* $\mathfrak{M}_{a,q}$*, then*

$$\int_{\mathfrak{M}} (T(\alpha))^s e(-N\alpha)d\alpha = P^{s-k}\mathfrak{S}(P^\delta, N)J(P^\delta) + O(P^{s-k-\delta'}) \qquad (4.8)$$

*for some* $\delta' > 0$*, where*

$$\mathfrak{S}(P^\delta, N) = \sum_{q \le P^\delta} \sum_{\substack{a=1 \\ (a,q)=1}}^{q} (q^{-1}S_{a,q})^s e(-Na/q), \qquad (4.9)$$

$$J(P^\delta) = \int_{|\gamma|<P^\delta} \left(\int_0^1 e(\gamma\xi^k)d\xi\right)^s e(-\gamma)d\gamma. \qquad (4.10)$$

*Proof.* We first raise to the power $s$ the expression (4.5) for $T(\alpha)$, valid on an individual major arc $\mathfrak{M}_{a,q}$. Since

$$|q^{-1}S_{a,q}I(\beta)| \le P$$

trivially, we get

$$(T(\alpha))^s = (q^{-1}S_{a,q})^s(I(\beta))^s + O(P^{s-1+2\delta}). \qquad (4.11)$$

Multiplying by $e(-N\alpha)$ and integrating over $\mathfrak{M}_{a,q}$, that is, over $|\beta| < P^{-k+\delta}$, the main term in the last expression gives

$$(q^{-1}S_{a,q})^s e(-Na/q) \int_{|\beta|<P^{-k+\delta}} (I(\beta))^s e(-N\beta)d\beta$$

The integral here is independent of $q$ and $a$, and therefore summation over $q$ and $a$ satisfying (4.3) gives

$$\mathfrak{S}(P^\delta, N) \int_{|\beta|<P^{-k+\delta}} (I(\beta))^s e(-N\beta)d\beta.$$

In the integrand we can replace $N$ by $P^k$ with a negligible error. Indeed we have $N - P^k \ll P^{k-1}$, so that

$$|e(-\beta N) - e(-\beta P^k)| \ll |\beta|P^{k-1} \ll P^{-1+\delta},$$

and the error in the integral is $\ll P^{-k+\delta}P^sP^{-1+\delta}$. Since a crude estimate for $|\mathfrak{S}(P^\delta, N)|$ is $P^{2\delta}$, this leads to a final error $P^{s-k-1+4\delta}$, which is negligible. The integral is now

$$\int_{|\beta|<P^{-k+\delta}} \left( \int_0^P e(\beta \boldsymbol{\xi}^k) d\boldsymbol{\xi} \right)^s e(-P^k\beta) d\beta,$$

and on putting $\xi = P\xi'$ and $\beta = P^{-k}\gamma$, this becomes

$$P^{s-k} J(P^\delta).$$

Thus we have obtained the main term in the result (4.8).

It remains to estimate the effect of the error term in (4.11). Integrated over $|\beta| < P^{-k+\delta}$, it becomes $\ll P^{s-k-1+3\delta}$. Summed over $a \leq q$ and over $q \leq P^\delta$, it becomes $P^{s-k-1+5\delta}$, and since $\delta$ is small this is of the form given in (4.8). □

**Definition.** Let

$$\mathfrak{S}(N) = \sum_{q=1}^{\infty} \sum_{\substack{a=1 \\ (a,q)=1}}^{q} (q^{-1} S_{a,q})^s e(-Na/q). \tag{4.12}$$

This is called the *singular series* for the problem of representing $N$ as a sum of $s$ positive integral $k$th powers. If $s \geq 2^k + 1$, the series is absolutely convergent, and uniformly with respect to $N$, for by the Corollary to Lemma 3.1 we have (with $K = 2^{k-1}$):

$$|(q^{-1} S_{a,q})^s e(-Na/q)| \ll q^{-s/K+\varepsilon} \ll q^{-2-1/K+\varepsilon}.$$

Later we shall prove that the same is true under the less restrictive condition that $s \geq 2k + 1$.

**Theorem 4.1.** *If $s \geq 2^k + 1$, the number $r(N)$ of representations of $N$ as a sum of $s$ positive integral $k$th powers satisfies*

$$r(N) = C_{k,s} N^{s/k-1} \mathfrak{S}(N) + O(N^{s/k-1-\delta'}) \tag{4.13}$$

*for some fixed $\delta' > 0$, where*

$$C_{k,s} = \frac{\Gamma(1+1/k)^s}{\Gamma(s/k)} > 0. \tag{4.14}$$

*Proof.* By (4.1) and Lemmas 4.1 and 4.3,

$$\begin{aligned} r(N) &= \left\{ \int_{\mathfrak{M}} + \int_{\mathfrak{m}} \right\} (T(\alpha))^s e(-N\alpha) d\alpha \\ &= P^{s-k} \mathfrak{S}(P^\delta, N) J(P^\delta) + O(P^{s-k-\delta'}). \end{aligned} \tag{4.15}$$

We first investigate the integral $J(P^\delta)$, defined in (4.10). The inner integral there can be expressed, by obvious changes of variable, in three ways:

$$\int_0^1 e(\gamma\boldsymbol{\xi}^k)d\boldsymbol{\xi} = k^{-1}\int_0^1 \zeta^{-1+1/k}e(\gamma\zeta)d\zeta = k^{-1}\gamma^{-1/k}\int_0^\gamma \zeta^{-1+1/k}e(\zeta)d\zeta,$$

where in the last expression we have supposed for simplicity that $\gamma$ is positive. The integral in the last expression is a bounded function of $\gamma$, by Dirichlet's test for the convergence of an infinite integral together with the fact that the integral is absolutely convergent at 0. Hence

$$\left|\int_0^1 e(\gamma\boldsymbol{\xi}^k)d\boldsymbol{\xi}\right| \ll |\gamma|^{-1/k}.$$

This enables us to extend to infinity the integration over $\gamma$ in (4.10); we obtain

$$J(P^\delta) = J + O(P^{-(s/k-1)\delta}),$$

where

$$J = \int_{-\infty}^{\infty}\left(k^{-1}\int_0^1 \zeta^{-1+1/k}e(\gamma\zeta)d\zeta\right)^s e(-\gamma)d\gamma. \tag{4.16}$$

Plainly $J$ depends only on $k$ and $s$, and we shall prove in a moment that $J = C_{k,s}$. We shall call $J$ the *singular integral* for the problem of representing $N$ as a sum of $s$ positive $k$th powers.

By the absolute convergence of the series $\mathfrak{S}(N)$ and the result just proved for $J(P^\delta)$, we can replace $\mathfrak{S}(P^\delta, N)$ in (4.15) by $\mathfrak{S}(N)$ and we can replace $J(P^\delta)$ by $J$, with errors which are permissible. We can also replace $P$ by $N^{1/k}$ with permissible error, and this gives (4.13), except for the proof that $J = C_{k,s}$. The exact value of $J$ is perhaps unimportant, but we need to know that $J > 0$.

To evaluate $J$ we start from the fact that

$$\int_{-\lambda}^{\lambda} e(\mu\gamma)d\gamma = \frac{\sin 2\pi\lambda\mu}{\pi\mu}.$$

Hence

$$\begin{aligned} k^s J &= \lim_{\lambda\to\infty}\int_0^1\cdots\int_0^1 (\zeta_1\cdots\zeta_s)^{-1+1/k}\frac{\sin 2\pi\lambda(\zeta_1+\cdots+\zeta_s-1)}{\pi(\zeta_1+\cdots+\zeta_s-1)}d\zeta_1\cdots d\zeta_s \\ &= \lim_{\lambda\to\infty}\int_0^s \phi(u)\frac{\sin 2\pi\lambda(u-1)}{\pi(u-1)}du, \end{aligned}$$

where

$$\phi(u) = \int_0^1 \cdots \int_0^1 \{\zeta_1 \cdots \zeta_{s-1}(u - \zeta_1 - \cdots - \zeta_{s-1})\}^{-1+1/k} \, d\zeta_1 \cdots d\zeta_{s-1},$$

and is taken over $\zeta_1, \ldots, \zeta_{s-1}$ for which $u-1 < \zeta_1 + \cdots + \zeta_{s-1} < u$. Here we have made a change of variable from $\zeta_s$ to $u$, where $\zeta_1 + \cdots + \zeta_s = u$.

We now recall Fourier's integral theorem for a finite interval, which states[1] that under certain conditions,

$$\lim_{\lambda \to \infty} \int_A^B \phi(u) \frac{\sin 2\pi\lambda(u - C)}{\pi(u - C)} du = \phi(C),$$

provided $A < C < B$. Assuming that this is applicable, we deduce that

$$\begin{aligned} k^s J &= \phi(1) \\ &= \int_0^1 \cdots \int_0^1 \{\zeta_1 \cdots \zeta_{s-1}(1 - \zeta_1 - \cdots - \zeta_{s-1})\}^{-1+1/k} \, d\zeta_1 \cdots d\zeta_{s-1}, \end{aligned}$$

where the integral is taken over $\zeta_1, \ldots, \zeta_{s-1}$ for which $0 < \zeta_1 + \cdots + \zeta_{s-1} < 1$. The last definite integral, over $s - 1$ variables, is an instance of an integral evaluated by Dirichlet; it is indeed an immediate extension of Euler's integral

$$B(p, q) = \int_0^1 x^{p-1}(1 - x)^{q-1} dx = \frac{\Gamma(p)\Gamma(q)}{\Gamma(p + q)}.$$

We have[2]

$$\phi(1) = \frac{\Gamma(1/k)^s}{\Gamma(s/k)},$$

whence

$$J = \left(\frac{1}{k}\right)^s \frac{\Gamma(1/k)^s}{\Gamma(s/k)} = \frac{\Gamma(1 + 1/k)^s}{\Gamma(s/k)}.$$

A sufficient condition for the validity of Fourier's integral theorem is that $\phi(u)$ should be of bounded variation. To verify this, put $\zeta_j = ut_j$. Then $\phi(u)$ is equal to

$$u^{s/k-1} \int_0^{1/u} \cdots \int_0^{1/u} \{t_1 \cdots t_{s-1}(1 - t_1 - \cdots - t_{s-1})\}^{-1+1/k} \, dt_1 \cdots dt_{s-1},$$

where the integral is over $t_1, \ldots, t_{s-1}$ for which $1 - 1/u < t_1 + \cdots + t_{s-1} < 1$. The region of integration contracts as $u$ increases, and the

[1] See [97, §9.43] for example.
[2] See [97, §12.5].

integrand does not involve $u$. Hence $\phi(u)$ is the product of $u^{-1+s/k}$ and a positive monotonic decreasing function of $u$, and is therefore a function of bounded variation. This completes the proof. □

**Note.** In our treatment of the singular integral, we have followed a paper of Landau [54]. For a slightly more general treatment, see a paper of Kestelman [52]. There are various devices by which the use of Fourier's integral theorem can be avoided; for example one can replace $I(\beta)$ by the finite sum

$$k^{-1} \sum_{0<m<P^k} m^{-1+1/k} e(\beta m),$$

or one can evaluate $J$ indirectly as in Vinogradov [93, Chapter 3]. But on the whole the reference to Fourier's integral theorem seems natural and appropriate.

In the asymptotic formula (4.13) one can regard the first factor, $C_{k,s}N^{s/k-1}$, as measuring the 'density' of the solutions of

$$x_1^k + \cdots + x_s^k = N, \quad x_1 > 0, \ldots, x_s > 0$$

in real numbers; it is the $(s-1)$-dimensional content of this portion of a hypersurface. Otherwise expressed, it is (with a negligible error) the $s$-dimensional volume of the region

$$N - \frac{1}{2} < x_1^k + \cdots + x_s^k < N + \frac{1}{2}, \quad x_1 > 0, \ldots, x_s > 0.$$

The second factor, $\mathfrak{S}(N)$, can be regarded as a compensating factor to allow for the fact that $k$th powers of integers are not as uniformly distributed as are $k$th powers of real numbers, in that they are constrained by congruence restrictions. (The relation between $\mathfrak{S}(N)$ and congruences will emerge in the next section.) Thus the conclusion we draw from the asymptotic formula, expressed in somewhat vague terms, is that asymptotically the representations of a large number as a sum of $s$ positive integral $k$th powers are actually dominated by these two influences, provided $s$ is greater than some function of $k$.

# 5

# Waring's problem: the singular series

We now study the singular series:

$$\mathfrak{S}(N) = \sum_{q=1}^{\infty} \sum_{\substack{a=1 \\ (a,q)=1}}^{q} \left(q^{-1}S_{a,q}\right)^{s} e(-aN/q). \tag{5.1}$$

We shall find that the value of $\mathfrak{S}(N)$ is closely related to the number of solutions of the congruences

$$x_1^k + \cdots + x_s^k \equiv N \pmod{q}$$

for all positive integers $q$, and indeed $\mathfrak{S}(N) = 0$ if any such congruence is insoluble. This might be expected from the appearance of the asymptotic formula, since then $r(N) = 0$.

We write

$$\mathfrak{S}(N) = \sum_{q=1}^{\infty} A(q), \quad A(q) = \sum_{\substack{a=1 \\ (a,q)=1}}^{q} \left(q^{-1}S_{a,q}\right)^{s} e(-aN/q). \tag{5.2}$$

**Lemma 5.1.** *If* $(q_1, q_2) = 1$, *then*

$$A(q_1 q_2) = A(q_1)A(q_2). \tag{5.3}$$

*Proof.* Write

$$f(a,q) = (S_{a,q})^{s} e(-aN/q).$$

We shall prove that if $(a_1, q_1) = (a_2, q_2) = 1$ and

$$\frac{a}{q} \equiv \frac{a_1}{q_1} + \frac{a_2}{q_2} \pmod{1}, \quad q = q_1 q_2, \tag{5.4}$$

then

$$f(a, q) = f(a_1, q_1) f(a_2, q_2). \tag{5.5}$$

This will suffice to give the result, for the relation (5.4) sets up a 1-to-1 correspondence between reduced residue classes $a \pmod q$ and pairs of reduced residue classes $a_1 \pmod{q_1}$ and $a_2 \pmod{q_2}$, whence

$$\sum_{\substack{a=1\\(a,q)=1}}^{q} f(a,q) = \left( \sum_{\substack{a_1=1\\(a_1,q_1)=1}}^{q_1} f(a_1,q_1) \right) \left( \sum_{\substack{a_2=1\\(a_2,q_2)=1}}^{q_2} f(a_2,q_2) \right).$$

To prove (5.5), we use a somewhat similar argument, but with complete sets of residues. Putting

$$\frac{z}{q} \equiv \frac{z_1}{q_1} + \frac{z_2}{q_2} \pmod 1,$$

we have

$$S_{a,q} = \sum_{z=1}^{q} e(az^k/q) = \sum_{z_1=1}^{q_1} \sum_{z_2=1}^{q_2} e\left( \frac{a}{q} q^k \left( \frac{z_1}{q_1} + \frac{z_2}{q_2} \right)^k \right),$$

and since

$$\frac{a}{q} q^k \left( \frac{z_1}{q_1} + \frac{z_2}{q_2} \right)^k \equiv \frac{a_1}{q_1} (q_2 z_1)^k + \frac{a_2}{q_2} (q_1 z_2)^k \pmod 1,$$

we get

$$\begin{aligned} S_{a,q} &= \sum_{z_1=1}^{q_1} e\left( \frac{a_1}{q_1} (q_2 z_1)^k \right) \sum_{z_2=1}^{q_2} e\left( \frac{a_2}{q_2} (q_1 z_2)^k \right) \\ &= S_{a_1,q_1} S_{a_2,q_2}. \end{aligned}$$

Since, in addition

$$e\left( -\frac{a}{q} N \right) = e\left( -\frac{a_1}{q_1} N \right) e\left( -\frac{a_2}{q_2} N \right),$$

we obtain (5.5). □

**Note.** This result of this lemma is in no way dependent on the fact that $S_{a,q}$ is formed with the special polynomial $z^k$. We could replace $z^k$

with any polynomial $f(z)$ with integral coefficients, and indeed we could replace $S_{a,q}$ with a multiple exponential sum

$$\sum e\left(\frac{a}{q}f(z_1,\ldots,z_n)\right),$$

where each of $z_1,\ldots,z_n$ runs through a complete set of residues (mod $q$). The proof is just the same.

**Lemma 5.2.** *If $s \geq 2^k+1$, we have*

$$\mathfrak{S}(N) = \prod_p \chi(p), \tag{5.6}$$

*where*

$$\chi(p) = 1 + \sum_{\nu=1}^{\infty} A(p^\nu). \tag{5.7}$$

*Also*

$$|\chi(p) - 1| \ll p^{-1-\delta} \tag{5.8}$$

*for some fixed $\delta > 0$.*

*Proof.* It follows from Lemma 5.1 that if $q = p_1^{\nu_1}p_2^{\nu_2}\cdots$ then

$$A(q) = A\left(p_1^{\nu_1}\right)A\left(p_2^{\nu_2}\right)\cdots.$$

Hence

$$\mathfrak{S}(N) = \sum_{q=1}^{\infty} A(q) = \prod_p \left\{\sum_{\nu=0}^{\infty} A(p^\nu)\right\} = \prod_p \chi(p),$$

and this is justified by the convergence of $\sum |A(q)|$, which was proved in the preceding chapter.

We already had the estimate

$$|A(q)| \ll q^{-1-1/K+\varepsilon} \ll q^{-1-\delta},$$

and this implies

$$|\chi(p) - 1| \ll \sum_{\nu=1}^{\infty} p^{-\nu(1+\delta)} \ll p^{-1-\delta},$$

which is (5.8). □

**Corollary.** *If $s \geq 2^k + 1$ there exists $p_0 = p_0(k)$ such that*

$$\frac{1}{2} \leq \prod_{p > p_0} \chi(p) \leq \frac{3}{2}.$$

This follows at once from (5.8), since we can take $\delta$ to depend on $k$ only. Again, we shall see that the result holds if $s \geq 2k + 1$.

**Definition.** Let $M(q)$ denote the number of solutions of the congruence

$$x_1^k + \cdots + x_s^k \equiv N \pmod{q},$$

with $0 < x_1, \ldots, x_s \leq q$.

**Lemma 5.3.** *We have*

$$1 + \sum_{\nu=1}^{n} A(p^\nu) = M(p^n)/p^{n(s-1)}, \tag{5.9}$$

*and consequently*

$$\chi(p) = \lim_{n \to \infty} M(p^n)/p^{n(s-1)}. \tag{5.10}$$

*Proof.* We can express $M(q)$ in terms of exponential sums by a procedure which can be regarded as an arithmetical analogue of that used to express $r(N)$ as an integral in (2.7). We have

$$M(q) = q^{-1} \sum_{t=1}^{q} \sum_{x_1=1}^{q} \cdots \sum_{x_s=1}^{q} e\left(\frac{t}{q}\left(x_1^k + \cdots + x_s^k - N\right)\right),$$

since the sum over $t$ gives $q$ if the congruence is satisfied and 0 otherwise. We collect together those values of $t$ which have the same highest common factor with $q$. If this highest common factor is denoted by $q/q_1$, the values of $t$ in question are $uq/q_1$, where $1 \leq u \leq q_1$ and $(u, q_1) = 1$. Hence

$$M(q) = q^{-1} \sum_{q_1 | q} \sum_{\substack{u=1 \\ (u,q_1)=1}}^{q_1} \sum_{x_1=1}^{q} \cdots \sum_{x_s=1}^{q} e\left(\frac{u}{q_1}\left(x_1^k + \cdots + x_s^k - N\right)\right).$$

Now

$$\sum_{x=1}^{q} e\left(\frac{u}{q_1} x^k\right) = \frac{q}{q_1} \sum_{x=1}^{q_1} e\left(\frac{u}{q_1} x^k\right) = \frac{q}{q_1} S_{u,q_1}.$$

Thus

$$\begin{aligned} M(q) &= q^{-1} \sum_{q_1 | q} \sum_{\substack{u=1 \\ (u,q_1)=1}}^{q_1} \left(\frac{q}{q_1}\right)^s (S_{u,q_1})^s e\left(-\frac{uN}{q_1}\right) \\ &= q^{s-1} \sum_{q_1 | q} A(q_1). \end{aligned}$$

This formula, when $q = p^n$, becomes (5.9), and (5.10) follows from it. □

**Note.** For each particular $N$, the series on the left of (5.9) terminates, and therefore (5.10) is true without the limiting operation for all sufficiently large $n$. However the point at which the series terminates depends on $N$, as well as on $k$ and $p$.

**Definition.** For each prime $p$, let $p^\tau$ be the highest power of $p$ dividing $k$, and put $k = p^\tau k_0$. Define $\gamma$ by

$$\gamma = \begin{cases} \tau + 1 & \text{if} \quad p > 2, \\ \tau + 2 & \text{if} \quad p = 2. \end{cases} \tag{5.11}$$

Of course, $\gamma$ depends on both $p$ and $k$.

**Lemma 5.4.** *If the congruence $y^k \equiv m \pmod{p^\gamma}$ is soluble where $m \not\equiv 0 \pmod{p}$, then the congruence $x^k \equiv m \pmod{p^\nu}$ is soluble for every $\nu > \gamma$.*

*Proof. Suppose $p > 2$.* The relatively prime residue classes (mod $p^\nu$) form a cyclic group of order $\phi(p^\nu) = p^{\nu-1}(p-1)$, being represented by the powers of a primitive root $g$ to the modulus $p^\nu$. If $\nu > \gamma$, then $g$ is necessarily also a primitive root to the modulus $p^\gamma$.

Write

$$m \equiv g^\mu, \quad y \equiv g^\eta, \quad x \equiv g^\xi \pmod{p^\nu}.$$

Then the hypothesis that $y^k \equiv m \pmod{p^\gamma}$ is equivalent to

$$k\eta \equiv \mu \pmod{p^{\gamma-1}(p-1)}.$$

Since $k = p^\tau k_0$ and $\tau = \gamma - 1$, it follows that $\mu$ is divisible by $p^{\gamma-1}$ and also by $(k_0, p-1)$. But now we can find $\xi$ to satisfy

$$k\xi \equiv \mu \pmod{p^{\nu-1}(p-1)},$$

since $\mu$ is divisible by the highest common factor of $k$ and $p^{\nu-1}(p-1)$. The last congruence is equivalent to $x^k \equiv m \pmod{p^\nu}$.

*Suppose* $p = 2$. First, if $\tau = 0$, so that $k$ is odd, there is no problem. For as $x$ runs through a reduced set of residues to the modulus $2^\nu$ then so does $x^k$, and the congruence $x^k \equiv m \pmod{2^\nu}$ is soluble for any odd $m$ without any hypothesis.

Now suppose $\tau \geq 1$. Since $k = 2^\tau k_0$ is even, we have $x^k \equiv 1 \pmod 4$ for all $x$. Those residue classes (mod $2^\nu$) that are $\equiv 1 \pmod 4$ constitute a cyclic group of order $2^{\nu-2}$, and it is well known that 5 is a generating element, i.e. a primitive root. As before, write

$$m \equiv 5^\mu, \quad y \equiv 5^\eta, \quad x \equiv 5^\xi \pmod{2^\nu}.$$

Then the hypothesis is equivalent to

$$k\eta \equiv \mu \pmod{2^{\gamma-2}}.$$

Since $k = 2^\tau k_0$ and $\tau = \gamma - 2$, it follows that $\mu$ is divisible by $2^\tau$. Hence there exists $\xi$ such that

$$k\xi \equiv \mu \pmod{2^{\nu-2}},$$

which implies that $x^k \equiv m \pmod{2^\nu}$. This completes the proof of Lemma 5.4. □

**Lemma 5.5.** *If the congruence*

$$x_1^k + \cdots + x_s^k \equiv N \pmod{p^\gamma}$$

*has a solution with $x_1, \ldots, x_s$ not all divisible by $p$, then*

$$\chi(p) > 0.$$

*Proof.* Suppose $a_1^k + \cdots + a_s^k \equiv N \pmod{p^\gamma}$ and $a_1 \not\equiv 0 \pmod p$. We can obtain many solutions of $x_1^k + \cdots + x_s^k \equiv N \pmod{p^\nu}$ for $\nu > \gamma$ by the following construction. We choose $x_2, \ldots, x_s$ arbitrarily, subject to

$$x_j \equiv a_j \pmod{p^\gamma}, \quad 0 < x_j \leq p^\nu.$$

These choices are possible in $p^{(\nu-\gamma)(s-1)}$ ways. Then choose $x_1$ to satisfy

$$x_1^k \equiv N - x_2^k - \cdots - x_s^k \pmod{p^\nu};$$

this is possible by Lemma 5.4 because the expression on the right is

$\equiv a_1^k \pmod{p^\nu}$ and $a_1 \not\equiv 0 \pmod{p}$. Thus, in the notation introduced earlier, we have

$$M(p^\nu) \geq p^{(\nu-\gamma)(s-1)} = C_p p^{\nu(s-1)},$$

where $C_p = p^{-\gamma(s-1)} > 0$. By (5.10) of Lemma 5.3, this implies $\chi(p) > 0$. □

**Lemma 5.6.** *If $s \geq 2k$ ($k$ odd) or $s \geq 4k$ ($k$ even), then $\chi(p) > 0$ for all primes $p$ and all $N$.*

*Proof.* By Lemma 5.5 it suffices to prove that the congruence

$$x_1^k + \cdots + x_s^k \equiv N \pmod{p^\gamma} \tag{5.12}$$

is soluble with $x_1, \ldots, x_s$ not all divisible by $p$. If $N \not\equiv 0 \pmod{p}$, the latter requirement is necessarily satisfied. If $N \equiv 0 \pmod{p}$, it will suffice to solve the congruence

$$x_1^k + \cdots + x_{s-1}^k + 1^k \equiv N \pmod{p^\gamma}.$$

Hence (replacing $s-1$ by $s$) we see that it suffices to prove the solubility of (5.12) when $N \not\equiv 0 \pmod{p}$ for $s \geq 2k-1$ ($k$ odd) or $s \geq 4k-1$ ($k$ even).

*Suppose* $p > 2$. We consider all $N$ satisfying

$$0 < N < p^\gamma, \quad N \not\equiv 0 \pmod{p},$$

their number being $\phi(p^\gamma) = p^{\gamma-1}(p-1)$. Let $s(N)$ denote the least $s$ for which (5.12) is soluble. If $N \equiv z^k N' \pmod{p^\gamma}$, then obviously $s(N) = s(N')$. Hence if we distribute the numbers $N$ into classes according to the value of $s(N)$, the number in each class is at least equal to the number of distinct values assumed by $z^k$ when $z \not\equiv 0 \pmod{p}$. By putting $z \equiv g^\varsigma \pmod{p^\gamma}$, where $g$ is a primitive root $\pmod{p^\gamma}$, and $a \equiv g^\alpha \pmod{p^\gamma}$, one easily sees that the congruence $z^k \equiv a \pmod{p^\gamma}$ is soluble if and only if $\alpha$ is divisible by $p^\tau \delta$ where $\delta = (k, p-1)$. Since $\tau = \gamma - 1$, the number of distinct values for $\alpha$ $\pmod{p^{\gamma-1}(p-1)}$, which is also the number of distinct values for $a$ $\pmod{p^\gamma}$, is

$$\frac{p^{\gamma-1}(p-1)}{p^{\gamma-1}\delta} = \frac{p-1}{\delta} = r,$$

say. Hence each class of values of $N$ includes at least $r$ elements.

Let us enumerate first all $N$ for which $s(N) = 1$:

$$N_1^{(1)} < N_2^{(1)} < \cdots < N_{r_1}^{(1)}, \quad \text{where } r_1 \geq r.$$

Then we enumerate all $N$ for which $s(N) = 2$:

$$N_1^{(2)} < N_2^{(2)} < \cdots < N_{r_2}^{(2)}, \quad \text{where } r_2 \geq r,$$

and so on. Some of these sets may be empty, but we shall prove that two consecutive sets cannot be empty.

Consider the least $N' \not\equiv 0 \pmod p$ which is not in any of the first $j-1$ sets. Then either $N'-1$ or $N'-2$ is $\not\equiv 0 \pmod p$, and being less than $N'$ it must be in one of the first $j-1$ sets. Representing $N'$ as

$$(N'-1) + 1^k \qquad \text{or} \qquad (N'-2) + 1^k + 1^k,$$

we deduce that $s(N') \leq j+1$. Hence the sets for which $s(N) = j$, $s(N) = j+1$ cannot both be empty.

Suppose the last set in the enumeration is that for which $s(N) = m$. Then at least $\frac{1}{2}(m-1)$ of the first $m-1$ sets are not empty, and also the $m$th set is not empty, making at least $\frac{1}{2}(m+1)$ non-empty sets. Since each set contains at least $r$ numbers, we have

$$\tfrac{1}{2}(m+1)r \leq \phi(p^\gamma) = p^{\gamma-1}(p-1),$$

whence

$$\begin{aligned}(m+1) \leq \frac{2p^{\gamma-1}(p-1)}{r} &= 2p^{\gamma-1}\delta \\ &= 2p^\tau(k_0, p-1) \leq 2k.\end{aligned}$$

Hence $m \leq 2k-1$, whence $s(N) \leq 2k-1$ for all $N$. Thus for $p > 2$, the congruence (5.12) is soluble for $s \geq 2k-1$.

*Suppose* $p = 2$. If $\tau = 0$, that is if $k$ is odd, the congruence (5.12) is soluble for $N \not\equiv 0 \pmod p$ when $s = 1$, as was remarked in the proof of Lemma 5.4. This proves the conclusion of Lemma 5.6 since then the only significant restriction on $s$ comes from the primes $p > 2$.

Now suppose $\tau \geq 1$, so that $k$ is even. We can suppose without loss of generality that $0 < N < 2^\gamma$, since $N$ is now odd. By taking all the $x_j$ in (5.12) to be 0 or 1, we can certainly solve the congruence if $s \geq 2^\gamma - 1$. Now

$$2^\gamma - 1 = 2^{\tau+2} - 1 \leq 4k - 1.$$

Hence it suffices if $s \geq 4k-1$, and this proves the conclusion of Lemma 5.6 in the case when $k$ is even. □

**Note.** Although the final argument might, at first sight, seem to be a crude one, we have in fact lost nothing if $k = 2^\tau$ and $\tau \geq 2$. For then

$x^{2^\tau} \equiv 1 \pmod{2^{\tau+2}}$ if $x$ is odd, and $x^{2^\tau} \equiv 0 \pmod{2^{\tau+2}}$ if $x$ is even, so the values of $x^k$ are in this case simply 0 and 1.

Hardy and Littlewood defined $\Gamma(k)$ to be the least value of $s$ such that the congruence (5.12) is soluble with $x_1, \ldots, x_s$ not all divisible by $p$, for all $p$ and all $N$. In this notation, Lemma 5.6 states that $\Gamma(k) \leq 2k$ when $k$ is odd and $\Gamma(k) \leq 4k$ when $k$ is even. Hardy and Littlewood made a very detailed study of $\Gamma(k)$ in P.N.VIII [41].[1] In particular they determined all types of $k$ for which $\Gamma(k) > k$. The first few values of $\Gamma(k)$ are

| $k$ | 3 | 4 | 5 | 6 | 7 | 8 | 9 | 10 | 11 | 12 | 13 | 14 | 15 | 16 |
|---|---|---|---|---|---|---|---|---|---|---|---|---|---|---|
| $\Gamma(k)$ | 4 | 16 | 5 | 9 | 4 | 32 | 13 | 12 | 11 | 16 | 6 | 14 | 15 | 64 |

**Theorem 5.1.** *If* $s \geq 2^k + 1$ *then*

$$\mathfrak{S}(N) \geq C_1(k, s) > 0$$

*for all* $N$.

*Proof.* The result follows from Lemma 5.6 and the Corollary to Lemma 5.2, since $2^k + 1 \geq 2k$ ($k$ odd) and $2^k + 1 \geq 4k$ ($k$ even, $k > 2$). □

Theorem 5.1 is a necessary supplement to Theorem 4.1, in that it shows that the main term in the asymptotic formula is $\gg N^{s/k-1}$, and that consequently, $r(N) \to \infty$ as $N \to \infty$.

In the present chapter I have followed for the most part Vinogradov's exposition [93, Chapter 2].[2] This is somewhat simpler than the original exposition of Hardy and Littlewood.

[1] See also Chowla [14].
[2] The case $p = 2$ of our Lemma 5.4 is inadvertently omitted by Vinogradov.

# 6

# The singular series continued

We now prove the result mentioned in connection with the Corollary to Lemma 3.1, namely that

$$|S_{a,q}| \ll q^{1-1/k}. \tag{6.1}$$

This implies that

$$|A(q)| \ll q^{1-s/k},$$

from which it follows that the singular series is absolutely convergent if $s \geq 2k+1$. Also (5.8) of Chapter 5 and the Corollary to Lemma 5.2, both hold under the same condition.

**Lemma 6.1.** *If* $a \not\equiv 0 \pmod{p}$ *and* $\delta = (k, p-1)$ *then*

$$|S_{a,p}| \leq (\delta - 1)p^{1/2}. \tag{6.2}$$

*Proof.* Since $x^k \equiv m \pmod{p}$ has the same number of solutions as $x^\delta \equiv m \pmod{p}$, we have

$$S_{a,p} = \sum_x e\left(\frac{a}{p}x^\delta\right).$$

Let $\chi$ be a primitive character (mod $p$) of order $\delta$. Then the number of solutions of $x^\delta \equiv t \pmod{p}$ is

$$1 + \chi(t) + \cdots + \chi^{\delta-1}(t).$$

Hence

$$S_{a.p} = \sum_t \{1 + \chi(t) + \cdots + \chi^{\delta-1}(t)\}\, e\left(\frac{a}{p}t\right), \tag{6.3}$$

where here (and elsewhere in this proof) summations are over a complete set of residues modulo $p$. The sum arising from the term 1 in the bracket is 0, since $a \not\equiv 0 \pmod p$.

If $\psi$ is any non-principal character (mod $p$), the sum

$$T(\psi) = \sum_t \psi(t) e\left(\frac{at}{p}\right)$$

is called a Gauss sum, to commemorate the important part played by such sums in Gauss's work on cyclotomy. We can easily prove that $|T(\psi)| = p^{1/2}$, as follows. We have

$$|T(\psi)|^2 = \sum_t \sum_u \psi(t)\overline{\psi}(u) e\left(\frac{a}{p}(t-u)\right).$$

Here we can omit $u = 0$, since $\psi(0) = 0$. Changing the variable from $t$ to $v$, where $t \equiv vu \pmod p$, we obtain

$$|T(\psi)|^2 = \sum_v \sum_{u \neq 0} \psi(v) e\left(\frac{au}{p}(v-1)\right).$$

The inner sum is $p-1$ if $v = 1$ and is $-\psi(v)$ otherwise. Hence

$$|T(\psi)|^2 = p\psi(1) - \sum_v \psi(v) = p.$$

This is the result stated earlier. Using this in (6.3) for $\psi = \chi, \ldots, \chi^{\delta-1}$ we obtain (6.2). □

**Note.** (6.2) remains valid if $p = 2$ (so that $\delta = 1$), but is then trivial since $a = 1$ and $S_{1,2} = 1 + e^{i\pi} = 0$.

**Lemma 6.2.** *Suppose $a \not\equiv 0 \pmod p$ and $p \nmid k$. Then, for $1 < \nu \leq k$,*

$$S_{a,p^\nu} = p^{\nu-1}, \tag{6.4}$$

*and for $\nu > k$,*

$$S_{a,p^\nu} = p^{k-1} S_{a,p^{\nu-k}}. \tag{6.5}$$

*Proof.* In the definition

$$S_{a,p^\nu} = \sum_{x=0}^{p^\nu - 1} e\left(\frac{a}{p^\nu} x^k\right),$$

we put $x = p^{\nu-1}y + z$ where $0 \leq y < p$, $0 \leq z < p^{\nu-1}$. Then

$$x^k \equiv z^k + kp^{\nu-1}z^{k-1}y \pmod{p^\nu},$$

since $2(\nu - 1) \geq \nu$. Hence

$$S_{a,p^\nu} = \sum_{z=0}^{p^{\nu-1}-1} \sum_{y=0}^{p-1} e\left(\frac{az^k}{p^\nu} + \frac{akz^{k-1}y}{p}\right).$$

Since $ak \not\equiv 0 \pmod p$, the inner sum is 0 unless $z \equiv 0 \pmod p$ in which case it is $p$. Hence, if $z = pw$

$$S_{a,p^\nu} = p \sum_{w=0}^{p^{\nu-2}-1} e\left(\frac{aw^k}{p^{\nu-k}}\right).$$

If $\nu \leq k$, all the terms in the last sum are 1, and we get $S_{a,p^\nu} = p^{\nu-1}$. If $\nu > k$, the general term is a periodic function of $w$ with period $p^{\nu-k}$, whence

$$S_{a,p^\nu} = pp^{k-2} S_{a,p^{\nu-k}}.$$

This proves the two results. (Note again that $p$ may be 2.) □

**Lemma 6.3.** *The second result of Lemma 6.2 holds also when $p \mid k$.*

*Proof.* Put $k = p^\tau k_0$, as earlier, and note that since $\nu > k$ we have

$$\nu > p^\tau k_0 \geq 2^\tau \geq \tau + 1,$$

whence $\nu \geq \tau + 2$. Indeed, $k \geq \tau + 2$, since $k \geq 6$ if $\tau = 1$.

We modify the previous proof by putting

$$x = p^{\nu-\tau-1} y + z, \quad 0 \leq y < p^{\tau+1}, \quad 0 \leq z < p^{\nu-\tau-1}.$$

We shall prove that

$$x^k \equiv z^k + kp^{\nu-\tau-1} z^{k-1} y \pmod p. \tag{6.6}$$

Assuming this, the proof can be completed as before. For then

$$S_{a,p^\nu} = \sum_{z=0}^{p^{\nu-\tau-1}-1} \sum_{y=0}^{p^{\tau+1}-1} e\left(\frac{az^k}{p^\nu} - \frac{ak_0 z^{k-1} y}{p}\right),$$

and again the inner sum is 0 unless $z \equiv 0 \pmod p$, whence

$$\begin{aligned} S_{a,p^\nu} &= p^{\tau+1} \sum_{w=0}^{p^{\nu-\tau-2}-1} e\left(\frac{aw^k}{p^{\nu-k}}\right) \\ &= p^{\tau+1} p^{k-\tau-2} S_{a,p^{\nu-k}}. \end{aligned}$$

This proves (6.5). It remains to prove the congruence (6.6). It will suffice to prove that

$$(z + p^{\nu-\tau-1}y)^{p^\tau} \equiv z^{p^\tau} + p^{\nu-1}z^{p^\tau-1}y \pmod{p^\nu}$$

since the further operation of raising both sides to the power $k_0$ presents no difficulty. Putting $\nu - \tau - 1 = \lambda$, we have to prove that

$$(z + p^\lambda y)^{p^\tau} \equiv z^{p^\tau} + p^{\lambda+\tau}z^{p^\tau-1}y \pmod{p^{\lambda+\tau+1}}. \tag{6.7}$$

This is not quite as immediate as it might appear, because not all the binomial coefficients in the expansion of $(A + B)^{p^\tau}$ are divisible by $p^\tau$. However, we can prove the result in stages (or in other words by induction on $\tau$). We prove first that

$$(z + p^\lambda y)^p \equiv z^p + p^{\lambda+1}z^{p-1}y \pmod{p^{\lambda+2}} \tag{6.8}$$

provided $\lambda \geq 1$ (if $p > 2$) or $\lambda \geq 2$ (if $p = 2$). The only term which needs examination in the binomial expansion is the last; for this we need $\lambda p \geq \lambda + 2$, and this is true if $\lambda \geq 1$ when $p > 2$, or if $\lambda \geq 2$ when $p = 2$.

Finally (6.7) follows by repetition from (6.8); at the next stage we obtain

$$\begin{aligned}(z + p^\lambda y)^{p^2} &= (z^p + p^{\lambda+1}z^{p-1}y_1)^p \\ &\equiv z^{p^2} + p^{\lambda+2}z^{p^2-1}y_1 \pmod{p^{\lambda+3}} \\ &\equiv z^{p^2} + p^{\lambda+2}z^{p^2-1}y \pmod{p^{\lambda+3}},\end{aligned}$$

where $y_1 \equiv y \pmod{p}$, and we have applied (6.8) with $\lambda + 1$ in place of $\lambda$. The argument continues, and gives (6.7).

The conditions on $\lambda$ are satisfied when $\lambda = \nu - \tau - 1$. We have already seen that $\nu - \tau - 1 \geq 1$, and if $p = 2$ we have $\nu \geq k + 1 \geq \tau + 3$, as noted earlier. Thus the proof is complete. □

**Lemma 6.4.** $|S_{a,q}| \ll q^{1-1/k}$ *for* $(a, q) = 1$.

*Proof.* Put $T(a, q) = q^{-1+1/k}S_{a,q}$. We have to prove that $T(a, q)$ is bounded independently of $q$. If $q = p_1^{\nu_1}p_2^{\nu_2}\cdots$, then by the multiplicative property of $S_{a,q}$ in the proof of Lemma 5.1, we have

$$T(a, q) = T(a_1, p_1^{\nu_1})T(a_2, p_2^{\nu_2})\cdots,$$

for suitable $a_1, a_2, \ldots$, each of which is relatively prime to the corresponding $p^\nu$. By the second part of Lemma 6.2 and Lemma 6.3, we have

$$T(a, p^\nu) = T(a, p^{\nu-k})$$

for $\nu > k$, so we can suppose all $v_i$ are $\leq k$.

By Lemma 6.1,

$$T(a,p) \leq kp^{1/2}p^{-(1-1/k)} \leq kp^{-1/6}$$

and by the first part of Lemma 6.2,

$$T(a,p^\nu) = p^{\nu-1}p^{-\nu(1-1/k)} \leq 1 \quad \text{for } 1 < \nu \leq k.$$

Hence $T(a,p^\nu) \leq 1$ except possibly if $\nu = 1$ and $p \leq k^6$. Hence

$$T(a,q) \leq \prod_{p \leq k^6} (kp^{-1/6}),$$

and the number on the right is independent of $q$. □

**Theorem 6.1.** *The singular series* $\mathfrak{S}(N)$ *and the product* $\prod_p \chi(p)$ *are absolutely convergent if* $s \geq 2k+1$ *and*

$$\mathfrak{S}(N) \geq C_1(k,s) > 0$$

*if* $s \geq 2k+1$ *(k odd) or* $s \geq 4k$ *(k even).*

*Proof.* The absolute convergence follows as before, using Lemma 6.4 in place of the Corollary to Lemma 3.1, and the final assertion follows from Lemma 5.6 and the Corollary to Lemma 5.2. □

Theorem 6.1 shows that there would be no difficulty in improving on the condition $s \geq 2^k + 1$ for the validity of the asymptotic formula, as far as the singular series alone is concerned (except when $k = 4$). The crux of the difficulty is with the minor arcs, and not with the singular series.

**Note.** Hardy and Littlewood proved that the singular series, in the form $\sum_{q=1}^{\infty} A(q)$, is absolutely convergent for $s \geq 4$, and the same applies to the product form. The essential idea is to make sure of the cancellation which occurs in the summation over $a$ in

$$\sum_{\substack{a=1 \\ (a,q)=1}}^{q} (q^{-1}S_{a,q})^s e\left(\frac{-Na}{q}\right).$$

The absolute convergence is, however, no longer uniform in $N$. If the absolute value of $q^{-1}S_{a,q}$ is taken, the condition $s \geq 2k+1$ of Theorem 6.1 is best possible.

It is an interesting question how the sum $\mathfrak{S}(N)$ of the singular series fluctuates with $N$. Each factor $\chi(p)$ depends mainly on the residue class (mod $p^\gamma$) to which $N$ belongs. The factors which fluctuate most as $N$ varies are those for which $p$ divides $k$, but those for which $p-1$ has a large factor in common with $k$ may also fluctuate appreciably.

In their early papers, Hardy and Littlewood worked mainly with the definition of $\mathfrak{S}(N)$ in terms of the exponential sums $S_{a,q}$, rather than with the expression in terms of congruences (mod $p^\nu$). In P.N. II [38] they had to prove that $\mathfrak{S}(N)$ has a positive lower bound in the case $k = 4$, $s = 21$. The factors $\chi(p)$ which fluctuate most as $N$ varies are in this case $\chi(2)$ and $\chi(5)$; the product of all the others does not differ appreciably from 1. They found that $\chi(5)$ varies between about 0.7 and 1.3. But $\chi(2)$ varies by a factor of about 200. Hardy and Littlewood showed that (in the particular case mentioned)

$$\begin{aligned}\chi(2) &= 1 - 1.3307\cos\frac{(2N-5)\pi}{16} + 0.415\cos\frac{(6N+1)\pi}{16}\\ &\quad -0.3793\cos\frac{(2N+3)\pi}{8} + \varepsilon(N),\end{aligned}$$

where $|\varepsilon(N)| < 0.002$. It can be verified that $\chi(2)$ becomes very small (but still positive) when $N \equiv 2$ or $3 \pmod{16}$. It is relatively large when $N \equiv 10$ or $11 \pmod{16}$. These results correspond to the fact that $x^4 \equiv 0$ or $1 \pmod{16}$, and that consequently the choices for $x_1, \ldots, x_{21}$ in $x_1^4 + \cdots + x_{21}^4 \equiv N \pmod{16}$ are much more restricted in the one case than in the other.

# 7

# The equation $c_1x_1^k + \cdots + c_sx_s^k = N$

We next consider the problem of representing a large positive integer $N$ in the form $c_1x_1^k + \cdots + c_sx_s^k$, where $c_1, \ldots, c_s$ are given positive integers and $x_1, \ldots, x_s$ are arbitrary positive integers. It is not true, without some further supposition, that every large $N$ is representable if $s \geq s_0(k)$ for some $s_0(k)$. For suppose that $c_1, \ldots, c_{s-1}$ are all divisible by some prime $p$ and that $c_s$ is not. Then an integer $N$, not divisible by $p$, can certainly not be representable if it does not have the same $k$th power character as $c_s$ to the modulus $p$.

We can obviously suppose, in treating the equation in the title, that $c_1, \ldots, c_s$ do not all have a common factor. We shall find it necessary to postulate, in order to ensure solubility, that the congruence

$$c_1x_1^k + \cdots + c_sx_s^k \equiv N \pmod{p^\nu}$$

is soluble for each prime $p$ and all sufficiently large $\nu$, with not all the terms $c_1x_1^k, \cdots, c_sx_s^k$ divisible by $p$ (or to make some other supposition from which this can be deduced).

Only slight changes are needed in the preceding work to adapt it to this more general equation. We define $P_j = [(N/c_j)^{1/k}]$ to be the integer part of $(N/c_j)^{1/k}$, and we define

$$T_j(\alpha) = \sum_{x=1}^{P_j} e(\alpha c_j x^k). \tag{7.1}$$

Weyl's inequality (Lemma 3.1) applies to the sum $T_j(\alpha)$; if $\alpha = a/q + \beta$ and $|\beta| < q^{-2}$ then $c_j\alpha = c_ja/q + c_j\beta$, and $|c_j\beta| \ll q^{-2}$. This is sufficient for the proof of Lemma 3.1, since all we used about $\beta$ was that $|\beta| \ll q^{-2}$.

Hua's inequality (Lemma 3.2) remains valid for any one sum, since

$$\begin{aligned}\int_0^1 |T_j(\alpha)|^{2^k}\, d\alpha &= \int_0^1 \left|\sum_{x=1}^{P_j} e(\alpha c_j x^k)\right|^{2^k} d\alpha \\ &= \frac{1}{c_j}\int_0^{c_j} \left|\sum_{x=1}^{P_j} e(\alpha x^k)\right|^{2^k} d\alpha \\ &= \int_0^1 \left|\sum_{x=1}^{P_j} e(\alpha x^k)\right|^{2^k} d\alpha \quad \text{(by periodicity)} \\ &\ll P_j^{2^k-k+\varepsilon}.\end{aligned}$$

This inequality also extends to any product of $2^k$ sums, by Hölder's inequality; we obtain

$$\int_0^1 |T_1(\alpha)\cdots T_{2^k}(\alpha)|\, d\alpha \ll (P_1\cdots P_{2^k})^{1-k/2^k+\varepsilon}. \tag{7.2}$$

We define major and minor arcs as before. Lemma 4.1 now states that if $s \geq 2^k + 1$, then

$$\int_{\mathfrak{m}} |T_1(\alpha)\cdots T_s(\alpha)|\, d\alpha \ll P^{s-k-\delta'},$$

the proof being as before, using (7.2). Lemma 4.2 is unchanged, except that $I(\beta)$ is replaced by

$$I_j(\beta) = \int_0^{P_j} e(\beta c_j \xi^k)\, d\xi.$$

It simplifies the later calculations slightly, however, if the upper limit is replaced by $Pc_j^{-1/k}$, where $P = N^{1/k}$; the difference is negligible.

The proofs of Lemma 4.3 and Theorem 4.1 apply to the present problem with only slight changes. One difference is that the change of variable which is made in $I_j(\beta)$ in order to express it in terms of $\int_0^1 e(\gamma\xi^k)\, d\xi$ produces a factor $|c_j|^{-1/k}$. Another difference is that the singular series is of a slightly more general form; it is now given by

$$\mathfrak{S}(N) = \sum_{q=1}^{\infty} \sum_{\substack{a=1 \\ (a,q)=1}}^{q} q^{-s} S_{c_1 a,q}\cdots S_{c_s a,q}\, e(-Na/q). \tag{7.3}$$

To establish the absolute convergence of this series for $s \geq 2^k + 1$ using the Corollary to Lemma 3.1, or for $s \geq 2k + 1$ using Lemma 6.4, we need to extend an estimate for $S_{a,q}$ when $(a, q) = 1$ so that it applies to $S_{ca,q}$, where $c$ is any fixed positive integer. This is an easy matter, for if $ca/q = a'/q'$ then

$$S_{ca,q} = \frac{q}{q'} S_{a',q'},$$

and $q/q'$, being a divisor of $c$, is bounded.

In this way we can prove the following more general form of Theorem 4.1:

**Theorem 7.1.** *Let* $c_1, \ldots, c_s$ *be fixed positive integers. Then if* $s \geq 2^k + 1$, *the number* $r(N)$ *of representations of* $N$ *as*

$$N = c_1x_1^k + \cdots + c_sx_s^k, \quad (x_1, \ldots, x_s > 0),$$

*satisfies*

$$r(N) = \frac{C_{k,s}}{(c_1c_2\cdots c_s)^{1/k}} N^{s/k-1}\mathfrak{S}(N) + O(N^{s/k-1-\delta}) \tag{7.4}$$

*for some fixed* $\delta > 0$, *where* $C_{k,s}$ *is as in Theorem 4.1, and* $\mathfrak{S}(N)$ *is defined by (7.3). The series (7.3) is absolutely convergent for* $s \geq 2k+1$.

Lemmas 5.1, 5.2, 5.3 on the factorization of the singular series and on the relation between the singular series and $M(p^\nu)$, still apply. Thus in order that $\mathfrak{S}(N)$ may have a positive lower bound independent of $N$, it suffices if, for each $p$, the number $M(p^\nu)$ of solutions of

$$c_1x_1^k + \cdots + c_sx_s^k \equiv N \pmod{p^\nu}$$

satisfies

$$M(p^\nu) \geq C_p p^{\nu(s-1)}$$

for all sufficiently large $\nu$.

Defining $\gamma$ as before, and using Lemma 5.4, we find, as in Lemma 5.5, that a sufficient condition for this is that the congruence

$$c_1x_1^k + \cdots + c_sx_s^k \equiv N \pmod{p^\nu} \tag{7.5}$$

shall have a solution with not all of $c_1x_1^k, \ldots, c_sx_s^k$ divisible by $p$. Hence:

**Theorem 7.2.** *Let* $\gamma$ *be defined by (5.11). Suppose that* $s \geq 2k + 1$, *and suppose that for each*[1] *prime $p$ the congruence (7.5) has a solution in which not all of $c_1x_1^k, \ldots, c_sx_s^k$ are divisible by $p$. Then for all $N$ satisfying this hypothesis, we have*

$$\mathfrak{S}(N) \geq C(k, s) > 0.$$

By Theorems 7.1 and 7.2, if $s \geq 2^k + 1$ then $r(N) \to \infty$ as $N \to \infty$, provided $N$ is restricted to numbers which satisfy the congruence condition of Theorem 7.2. Since the congruence condition is needed only for $p \leq p_0$, and since $\gamma$ is independent of $N$, the numbers $N$ which satisfy the congruence condition will certainly include all numbers in some arithmetic progression.

If we make the hypothesis that the coefficients $c_1, \ldots, c_s$ are relatively prime in pairs, we can show that the congruence condition is satisfied for all $N$ provided $s$ exceeds some specific function of $k$. We prove:

**Theorem 7.3.** *Suppose that $(c_i, c_j) = 1$ for $1 \leq i < j \leq s$, and suppose that $s \geq k(2k-1)+2$ ($k$ odd) or $s \geq 2k(4k-1)+2$ ($k$ even). Then*

$$\mathfrak{S}(N) \geq C(k, s) > 0.$$

*Proof.* We have to prove that, under the conditions stated, the congruence (7.5) has a solution with not all of $c_1x_1^k, \ldots, c_sx_s^k$ divisible by $p$. Since at most one of $c_1, \ldots, c_s$ can be divisible by $p$, it will be enough to solve

$$c_1x_1^k + \cdots + c_sx_s^k \equiv N \pmod{p^\gamma} \tag{7.6}$$

with not all of $x_1, \ldots, x_{s-1}$ divisible by $p$, on the supposition that none of $c_1, \ldots, c_{s-1}$ is divisible by $p$.

*Suppose* $p > 2$. We saw in the proof of Lemma 5.6 that the number of distinct values assumed by $z^k$ to the modulus $p^\gamma$, when $z \not\equiv 0 \pmod p$, is $(p-1)/\delta$, where $\delta = (k_0, p-1)$. Hence the number of different classes of $k$th power residues and non-residues (mod $p^\gamma$) is

$$\frac{\phi(p^\gamma)}{(p-1)/\delta} = \frac{p^{\gamma-1}(p-1)\delta}{p-1} = p^{\gamma-1}\delta.$$

(These classes are the cosets of the subgroup of $k$th powers in the whole group of the relatively prime residue classes.)

[1] It suffices, of course, to suppose this for $p \leq p_0(k, s)$; see the Corollary to Lemma 5.2.

If we divide the coefficients $c_1, \ldots, c_{s-1}$ into sets according to the class of $k$th power residues or non-residues to which a coefficient belongs, there will be one class containing at least $(s-1)/p^{\gamma-1}\delta$ coefficients. Let $t$ be the least integer $\geq (s-1)/p^{\gamma-1}\delta$. We can take the coefficients in question to be the first $t$ coefficients, and then $c_2 \equiv d_2^k c_1, \ldots, c_t \equiv d_t^k c_1 \pmod{p^\gamma}$, where the $d_i$ are not divisible by $p$. Putting the variables $x_{t+1}, \ldots$ in (7.6) equal to 0, and cancelling $c_1$, we see that it suffices to solve

$$x_1^k + (d_2x_2)^k + \cdots + (d_tx_t)^k \equiv N' \pmod{p^\gamma}$$

with not all the variables divisible by $p$. This is in effect the same as the congruence considered in Lemma 5.6 in connection with Waring's problem. We proved there that the result holds provided $t \geq 2k$. Hence it suffices if

$$\frac{s-1}{p^{\gamma-1}\delta} > 2k - 1.$$

Since $\gamma - 1 = \tau$ and $p^\tau\delta \leq p^\tau k_0 = k$, it suffices if $s - 1 > k(2k-1)$.

*Suppose* $p = 2$. First, if $\tau = 0$ (so that $k$ is odd), the congruence (7.6) is soluble even if it has only one term provided $N$ is odd, since $x^k$ assumes all values $(\text{mod } p^\gamma)$. Hence it is soluble with two terms whether $N$ is odd or even, so it suffices if $s - 1 \geq 2$. Thus the conclusion of the theorem holds if $k$ is odd.

Now suppose $\tau \geq 1$, so that $k$ is even. Since each coefficient $c_i$ is odd (for $i \leq s-1$), it can assume $2^{\gamma-1}$ possible values to the modulus $2^\gamma$. Hence there is some set of $t$ mutually congruent coefficients, where $t \geq (s-1)/2^{\gamma-1}$. Putting the variables corresponding to the other coefficients equal to 0, we see that it suffices to solve

$$x_1^k + \cdots + x_t^k \equiv N' \pmod{2^\gamma}$$

with not all the variables even. As in the proof of Lemma 5.6, it suffices if $t \geq 4k$. Hence it suffices if

$$\frac{s-1}{2^{\gamma-1}} > 4k - 1.$$

Since $k \geq 2^\tau$ and $\gamma = \tau + 2$, we have $2^{\gamma-1} \leq 2k$. Hence it suffices if $s - 1 > 2k(4k-1)$. This proves Theorem 7.3 in the case when $k$ is even.

□

It follows from Theorems 7.1 and 7.3 that we can name a number $s_1(k)$ such that if $s \geq s_1(k)$ then $r(N) \to \infty$ as $N \to \infty$; always on the assumption that the coefficients $c_j$ are relatively prime in pairs. The numbers given in Theorem 7.3 are by no means best possible; we have merely given those which turn up naturally from the simple line of argument used in the proof. In principle, one can relax the condition that the coefficients are relatively prime in pairs; what is essential for the truth of the result just stated is that, for any prime $p$, a certain number of the coefficients are not divisible by $p$.

# 8

# The equation $c_1 x_1^k + \cdots + c_s x_s^k = 0$

We now study the solutions of the above equation in integers, positive or negative, where $c_1, \ldots, c_s$ are fixed integers, none of them 0. If $k$ is even, we must obviously suppose that not all the coefficients are of the same sign. If $k$ is odd, we can ensure this by changing $x_i$ into $-x_i$ if necessary. Hence, with a slight change of notation, we can write the equation as

$$c_1 x_1^k + \cdots + c_r x_r^k - c_{r+1} x_{r+1}^k - \cdots - c_s x_s^k = 0, \tag{8.1}$$

where $c_1, \ldots, c_s$ are now positive integers and $1 \leq r \leq s-1$. We study the solutions of (8.1) in *positive* integers.

The first difference, in comparison with the equation treated in Chapter 7, is that there is no large number $N$ which imposes restrictions on the sizes of the unknowns. We must therefore ourselves prescribe ranges for the variables, and the obvious way to do this is to choose a large number $P$, define $P_j = [P/c_j^{1/k}]$ for $1 \leq j \leq s$, and consider the number of solutions of (8.1) subject to

$$1 \leq x_j \leq P_j, \quad (1 \leq j \leq s). \tag{8.2}$$

We define the exponential sums $T_j(\alpha)$ as before, in (7.1). Then the number $\mathcal{N}(P)$ of solutions of (8.1), subject to (8.2), is given by

$$\mathcal{N}(P) = \int_0^1 T_1(\alpha) \cdots T_r(\alpha) T_{r+1}(-\alpha) \cdots T_s(-\alpha) d\alpha.$$

We follow again the treatment of Waring's problem, with the same slight changes as in the preceding section. The only further changes arise from the absence of $N$ in the singular series and in the singular integral. In (4.10), we have to replace $J(P^\delta)$ by

$$(c_1 \cdots c_s)^{-1/k} \int_{|\gamma|<P^\delta} \left( \prod_{j=1}^{s} \int_0^1 e(\pm\gamma\xi_j^k) d\xi_j \right) d\gamma,$$

where the sign is $+$ for $j \leq r$ and $-$ for $j > r$. As in (4.16) we are led to the evaluation of the integral

$$J = \int_{-\infty}^{\infty} k^{-s} \left( \prod_{j=1}^{s} \int_0^1 \zeta_j^{-1+1/k} e(\pm\gamma\zeta_j) d\zeta_j \right) d\gamma.$$

As in the proof of Theorem 4.1, we make a change of variable from $\zeta_s$ to $u$, where

$$\zeta_1 + \cdots + \zeta_r - \zeta_{r+1} - \cdots - \zeta_s = u,$$

and we find that

$$J = k^{-s} \int_0^1 \cdots \int_0^1 \{\zeta_1 \cdots \zeta_s(\zeta_1 \pm \cdots \pm \zeta_{s-1})\}^{-1+1/k} d\zeta_1 \cdots d\zeta_{s-1},$$

where $0 < \zeta_1 \pm \cdots \pm \zeta_{s-1} < 1$. All we need to know is that $J > 0$, and this is the case because there is some open set contained in $0 < \zeta_j < 1$ throughout which

$$0 < \zeta_1 \pm \cdots \pm \zeta_{s-1} < 1.$$

Thus the asymptotic formula for $\mathcal{N}(P)$, proved for $s \geq 2^k + 1$, takes the form

$$\mathcal{N}(P) = \frac{C'_{k,s}}{(c_1 \cdots c_s)^{1/k}} P^{s-k} \mathfrak{S} + O(P^{s-k-\delta}), \tag{8.3}$$

where

$$C'_{k,s} = k^{-s} \int_0^1 \cdots \int_0^1 \{\eta_1 \cdots \eta_s(\eta_1 + \cdots - \eta_{s-1})\}^{-1+1/k} d\eta_1 \cdots d\eta_{s-1}, \tag{8.4}$$

and

$$\mathfrak{S} = \sum_{q=1}^{\infty} \sum_{\substack{a=1 \\ (a,q)=1}}^{q} q^{-s} S_{c_1 a,q} \cdots S_{-c_s a,q}. \tag{8.5}$$

We observe that the value of $\mathfrak{S}$ is now a number depending only on the coefficients $c_1, \ldots, -c_s$ and on $k$. As before, the series defining $\mathfrak{S}$ is absolutely convergent for $s \geq 2k+1$, and factorizes as $\prod \chi(p)$. Again there exists $p_0$ such that

$$\prod_{p>p_0} \chi(p) \geq \frac{1}{2}.$$

To ensure that $\mathfrak{S} > 0$ (there is now no need to write $\mathfrak{S} \geq C_1(k,s) > 0$, since there is no parameter $N$), it will suffice to prove that $\chi_p > 0$ for each individual $p$. As before, it suffices if

$$M(p^\nu) \geq C_p p^{\nu(s-1)} \tag{8.6}$$

for all sufficiently large $\nu$, where $M(p^\nu)$ denotes the total number of solutions of the congruence

$$c_1x_1^k + \cdots - c_sx_s^k \equiv 0 \pmod{p^\nu}, \qquad 0 \leq x < p^\nu. \tag{8.7}$$

Our object now is to obtain some explicit function $s_1(k)$ of $k$, such that (8.6) holds for each $p$ if $s \geq s_1(k)$. Then the asymptotic formula (8.3) will be significant, in the sense that the main term will be $\gg P^{s-k}$, and will imply that $\mathcal{N}(P) \to \infty$ as $P \to \infty$. In proving this result, the signs of the coefficients in (8.7) play no part, and therefore we revert to the original notation, in which there were no negative signs prefixed to the coefficients.

The first step is to derive a congruence in a smaller number of unknowns in which none of the coefficients are divisible by $p$, which is such that if (8.6) holds for the new congruence then it holds for the original congruence. We write

$$c_j = d_j p^{h_j k + l_j}, \quad (1 \leq j \leq s),$$

where

$$p \nmid d_j, \quad 0 \leq l_j < k.$$

Then (8.7) becomes

$$\sum_{j=1}^{s} d_j p^{l_j} \left(p^{h_j} x_j\right)^k \equiv 0 \pmod{p^\nu}.$$

Let $h = \max h_j$. We restrict ourselves to solutions of the form

$$x_j = p^{h-h_j} y_j.$$

Thus, for large $\nu$, we can cancel $p^{hk}$ from the congruence, and it becomes

$$\sum_{j=1}^{s} d_j p^{l_j} y_j^k \equiv 0 \pmod{p^{\nu-hk}}, \tag{8.8}$$

subject to

$$0 \leq y < p^{\nu - h + h_j}.$$

If we denote by $M'(p^{\nu-hk})$ the number of solutions of (8.8) subject to

$$0 \leq y < p^{\nu - hk}$$

then (since $h - h_j < hk$) we have

$$M(p^\nu) \geq M'(p^{\nu-hk}).$$

Hence it suffices to prove the analogue of (8.6) for $M'(p^\nu)$.

Let $l = \max l_j$. In the new congruence (8.8), but to the modulus $p^\nu$, we group together the terms according to the value of $l_j$. There are $k$ groups, and one at least of these must contain $v$ terms, where $v \geq s/k$. We put $y_j = py_j'$ in the other terms, and after dividing out a factor $p^l$ we obtain a congruence of the form

$$d_1 y_1^k + \cdots + d_v y_v^k + p(d_{v+1} y_{v+1}^k + \cdots) + \cdots \equiv 0 \pmod{p^{\nu - l}}. \quad (8.9)$$

Again we can replace $\nu - l$ on the right by $\nu$, since this merely changes $C_p$ in a result of the type (8.6). In the last congruence, we have

$$d_1 d_2 \cdots d_v \not\equiv 0 \pmod{p}.$$

Define $\gamma$ as usual (see (5.11)). By the argument used in the proof of Lemma 5.5, the desired result (8.6) will hold for the congruence (8.9), provided the congruence

$$d_1 y_1^k + \cdots + d_v y_v^k \equiv 0 \pmod{p^\gamma} \quad (8.10)$$

has a solution in which $y_1, \ldots, y_v$ are not all divisible by $p$.

*Suppose* $p > 2$. We argue as in the proof of Theorem 7.2, dividing the terms in (8.10) into groups according to the class of the $k$th power residues or non-residues to which the coefficient $d_j$ belongs. It suffices if

$$\frac{v}{p^{\gamma-1}\delta} > 2k - 1,$$

and since $p^{\gamma-1}\delta = p^\tau \delta \leq k$, it suffices if

$$v > k(2k-1).$$

Hence it suffices if

$$s > k^2(2k-1).$$

*Suppose* $p = 2$. Once again there is no problem if $\tau = 0$, so that $k$ is odd. If $\tau > 0$, we could argue as in the proof of Theorem 7.2, but there is a more effective argument which is quite simple. We shall prove that

$$d_1y_1^k + \cdots + d_vy_v^k \equiv 0 \pmod{2^\gamma}$$

has a solution with $y_1, \ldots, y_v$ not all even provided $v \geq 2^\gamma$. We find these solutions by taking $y_j = 0$ or $1$ (this being no loss of generality when $k$ is a power of 2, as remarked in connection with Lemma 5.6).

First, if $\gamma = 1$, we can solve

$$d_1t_1 + d_2t_2 \equiv 0 \pmod{2}$$

by taking $t_1 = t_2 = 1$ (since $d_1$, $d_2$ are odd). Next, we can solve

$$d_1t_1 + d_2t_2 + d_3t_3 + d_4t_4 \equiv 0 \pmod{4}$$

by taking either $t_1 = t_2 = 1$, $t_3 = t_4 = 0$ (if $d_1 + d_2 \equiv 0 \pmod{4}$) or $t_1 = t_2 = 0$, $t_3 = t_4 = 1$ (if $d_3 + d_4 \equiv 0 \pmod{4}$) or $t_1 = t_2 = t_3 = t_4 = 1$. The process continues, and the proof is easily completed by induction on $\gamma$.

The condition $v \geq 2^\gamma$ is satisfied if $v \geq 4k$, and therefore is satisfied if $s > k(4k-1)$. Since this number is less than $k^2(2k-1)$, it has no effect in the result.

Collecting our results, we have proved:

**Theorem 8.1.** *Let $c_1, \ldots, c_s$ be given integers, none of them 0, and not all of the same sign if $k$ is even. Then provided $s \geq 2^k + 1$, and*

$$s \geq k^2(2k-1) + 1, \tag{8.11}$$

*the equation*

$$c_1x_1^k + \cdots + c_sx_s^k = 0$$

*has infinitely many solutions in integers $x_1, \ldots x_s$, none of them 0.*

The condition in (8.11), which came from our investigation of the singular series, is not best possible. In [27] Davenport and Lewis show that to ensure $\mathfrak{S} > 0$, it suffices if

$$s \geq k^2 + 1.$$

This condition is best possible if $k+1$ is a prime $p$. For then $x^k \equiv 1 \pmod{p}$ if $x \not\equiv 0 \pmod{p}$, and it is easily deduced that the congruence

$$p^k | \left((x_1^k + \cdots + x_k^k) + p(x_{k+1}^k + \cdots + x_{2k}^k) + \cdots + p^{k-1} \times (x_{k^2-k+1}^k + \cdots + x_{k^2}^k)\right),$$

in $k^2$ variables, is insoluble unless all the variables are divisible by $p$. However, for most values of $k$ a smaller value than $k^2 + 1$ will suffice.

In the preceding treatment of the equation

$$c_1 x_1^k + \cdots + c_s x_s^k = 0$$

we have obtained an asymptotic formula for the number of integer solutions in the $s$-dimensional box $0 < x_j \leq P_j$ as $P \to \infty$. But this box is not related in any unique way to the equation, and the interest of the result lies mainly in the fact that it establishes the existence of an infinity of solutions. To prove this, however, it is not essential to obtain an asymptotic formula for *all* solutions in such a box; it would be enough to consider some special subset. Thus we can use methods similar to those developed for the estimation of $G(k)$ in Waring's problem. In Chapter 9 we shall study Vinogradov's method, which is very effective for large $k$, and in the subsequent chapter we shall adapt this method to the equation which we have been studying.

It should not be overlooked, however, that the method which we have been using is particularly appropriate to the study of the *distribution* of the solutions of the equation. Suppose $\lambda_1, \ldots, \lambda_s$ are any real numbers, none of them 0, which satisfy the equation

$$c_1 \lambda_1^k + \cdots + c_s \lambda_s^k = 0.$$

Then the method of the present section enables one to find an asymptotic formula, as $P \to \infty$, for the integral solutions of our equation in the box

$$1 - \delta < \frac{x_j}{\lambda_j P} < 1 + \delta,$$

for any small fixed positive number $\delta$. Expressed geometrically, the result means that the 'rays' from the origin to the integer points on the cone

$$c_1 x_1^k + \cdots + c_s x_s^k = 0$$

are everywhere dense on this cone, if the cone is considered as a real locus in $s$-dimensional space. Thus, although the method which is now to be expounded is more effective in establishing an infinity of solutions, it does not entirely supersede the previous method.

# 9

# Waring's problem: the number $G(k)$

The number $G(k)$ was defined (by Hardy and Littlewood) to be the least value of $s$ with the property that every sufficiently large integer $N$ is representable as a sum of $s$ positive integral $k$th powers. We already know, from Theorems 4.1 and 5.1, that $G(k) \leq 2^k + 1$. In the opposite direction, it is easily deduced from considerations of density that $G(k) \geq k + 1$; in fact the number of sets of integers $x_1, \ldots, x_k$ satisfying

$$x_1^k + \cdots + x_k^k \leq X, \quad 0 < x_1 \leq x_2 \leq \cdots \leq x_k$$

is easily seen to be asymptotic to $\gamma X$ as $X \to \infty$, where $\gamma < 1$ (by comparison with a multiple integral), and consequently there are many numbers not representable by $k$ $k$th powers. A better lower bound is often deducible from congruence conditions; we have $G(k) \geq \Gamma(k)$, where $\Gamma(k)$ is the number defined in Chapter 4.

There are better upper bounds available for $G(k)$ when $k$ is large. In 1934 Vinogradov proved that $G(k) < (6 + \delta)k \log k$ for $k > k_0(\delta)$, where $\delta$ is any small positive number [92]. We shall now give an exposition of the proof.[1] The numerical coefficient 6 was subsequently improved to 3 by Vinogradov in 1947 [93], but the proof is somewhat more difficult.

It will be recalled that in Chapter 4 we divided the range of integration for $\alpha$ into major arcs and minor arcs, the major arcs comprising those $\alpha$ that admit a rational approximation $a/q$ with

$$q \leq P^\delta, \quad |\alpha - a/q| < P^{-k+\delta}.$$

Compared with what is usually needed in work on Waring's problem, these major arcs were exceptionally few in number and short in length. It was possible to make the choice because of the very effective estimate of Hua (Lemma 3.2), which 'saves' almost $P^k$ in the estimation of

[1] We base this mainly on Heilbronn's account [44].

$\int |T(\alpha)|^{2^k}\, d\alpha$. In the present treatment we cannot include as much of the integral of $\alpha$ in the minor arcs. We therefore need a more effective method of approximation to $T(\alpha)$ on the major arcs than the very crude one used in the proof of Lemma 4.2. The question is essentially one of replacing a sum by an integral, and we shall use the following lemma of van der Corput, which is of independent interest.

**Lemma 9.1. (van der Corput)** *Suppose $f(x)$ is a real function which is twice differentiable for $A \le x \le B$. Suppose further that, in this interval,*

$$0 \le f'(x) \le \tfrac{1}{2}, \quad f''(x) \ge 0.$$

*Then*

$$\sum_{A \le n \le B} e(f(n)) = \int_A^B e(f(x))\, dx + O(1).$$

*Proof.* It will suffice to prove the result when $A$, $B$ are integers with $A < B$, and when the end terms $n = A$, $n = B$ in the sum are counted with factors $\frac{1}{2}$. By replacing $f(x)$ with $f(x) + c$, which is equivalent to multiplying both the sum and the integral by $e^{2\pi i c}$, we can ensure that the difference between the sum and the integral is real, and this allows us to replace $e(f(x))$ by $\cos(2\pi f(x))$ on both sides.

Let $\Psi(x) = x - [x] - \frac{1}{2}$. Then, for any integer $m$ and any differentiable $F(x)$, we hae

$$\int_m^{m+1} \Psi(x) F'(x)\, dx = \tfrac{1}{2}\{F(m+1) + F(m)\} - \int_m^{m+1} F(x)\, dx.$$

Summing this for $m = A, A+1, \ldots, B-1$, we obtain

$$\sum_{n=A}^{B}{}' F(n) = \int_A^B F(x)\, dx + \int_A^B \Psi(x) F'(x)\, dx,$$

where the accent means that the end terms are counted with factors $\frac{1}{2}$. Thus the question is reduced to proving that

$$I = \int_A^B \Psi(x) \left(\cos(2\pi f(x))\right)'\, dx$$

is bounded in absolute value.

We recall that, for any $x$ which is not an integer,

$$\Psi(x) = -\sum_{\nu=1}^{\infty} \frac{\sin 2\pi \nu x}{\pi \nu}.$$

Hence

$$\begin{aligned} I &= -\sum_{\nu=1}^{\infty} \frac{1}{\nu\pi} \int_A^B (\sin 2\pi\nu x)\,(\cos(2\pi f(x)))'\ dx \\ &= 2\sum_{\nu=1}^{\infty} \frac{1}{\nu} \int_A^B (\sin 2\pi\nu x)\,(\sin(2\pi f(x)))\, f'(x)\ dx \\ &= \sum_{\nu=1}^{\infty} \frac{1}{\nu} \int_A^B f'(x)\left\{\cos\left(2\pi\left(\nu x - f(x)\right)\right) - \cos\left(2\pi\left(\nu x + f(x)\right)\right)\right\}\ dx. \end{aligned}$$

The interchange of summation and integration on the first line is easily justified by appealing to the bounded convergence of the series for $\Psi(x)$. We shall prove that

$$\left| \int_A^B f'(x) \cos\left(2\pi\left(\nu x \pm f(x)\right)\right)\ dx \right| < \frac{1}{\pi(2\nu - 1)}$$

and this will imply

$$|I| < \frac{1}{\pi} \sum_{\nu=1}^{\infty} \frac{1}{\nu(2\nu-1)} < \frac{2}{\pi},$$

giving the desired result.

We write the integral as

$$\frac{1}{2\pi} \int_A^B \frac{f'(x)}{\nu \pm f'(x)} \phi'(x)\ dx,$$

where

$$\phi(x) = \sin(2\pi(\nu x \pm f(x))),$$

and appeal to the mean value theorem. The second factor, $\phi'(x)$, has the property that its integral between any two limits has absolute value at most 2. The first factor is monotonic (for each positive integer $\nu$), its derivative being

$$\frac{\nu f''(x)}{(\nu \pm f'(x))^2} \geq 0.$$

The maximum of the first factor is at most $1/(2\nu - 1)$. Hence

$$\left| \frac{1}{2\pi} \int_A^B \frac{f'(x)}{\nu \pm f'(x)} \phi'(x)\ dx \right| \leq \frac{1}{\pi(2\nu - 1)},$$

as asserted above. Thus the proof of Lemma 9.1 is complete. □

We define the major arcs $\mathfrak{M}_{a,q}$ for the purpose of the present chapter, to consist of the intervals

$$(a,q)=1, \quad 1 \le a \le q, \quad q \le P^{1/2}, \quad |q\alpha - a| < \frac{1}{2kP^{k-1}}. \tag{9.1}$$

As in Chapter 2, we define

$$T(\alpha) = \sum_{x=1}^{P} e(\alpha x^k).$$

The more precise result, which takes the place of Lemma 4.2, is as follows.

**Lemma 9.2.** *For $\alpha$ in $\mathfrak{M}_{a,q}$, we have*

$$T(\alpha) = q^{-1} S_{a,q} I(\beta) + O(q), \tag{9.2}$$

*with the notation of Chapter 4.*

*Proof.* Putting $x = qy + z$, as in the proof of Lemma 4.2, we obtain

$$T(\alpha) = \sum_{z=1}^{q} e(az^k/q) \sum_{y} e\left(\beta(qy+z)^k\right),$$

the summation for $y$ being over $0 < qy + z \le P$. If

$$f(y) = \beta(qy+z)^k,$$

then (for $\beta > 0$),

$$f'(y) = k\beta q(qy+z)^{k-1} < k(2kP^{k-1})^{-1}P^{k-1} = \tfrac{1}{2},$$

by (9.1). Also $f''(y) \ge 0$. Hence Lemma 9.1 is applicable, and is equally applicable if $\beta < 0$ to the complex conjugate sum. Hence we can replace the inner sum over $y$ by

$$\int e\left(\beta(q\eta+z)^k\right)\, d\eta + O(1),$$

and this leads to (9.2). □

**Note.** It will be seen that the condition $q \le P^{1/2}$ in (9.1) has not been used in the proof, though of course the result loses its value if $q$ is allowed to be almost as large as $P$.

There is a still more effective method of approximating to $T(\alpha)$. If it is assumed that $q \le P^{1-\delta}$ and $|q\alpha - a| < P^{-k+1-\delta}$, then[1]

$$T(\alpha) = q^{-1} S_{a,q} I(\beta) + O\left(q^{\frac{3}{4}+\varepsilon}\right).$$

It is remarkable that the exponent in the error term here should be independent of $k$.

**Lemma 9.3.** *Suppose $s \ge 4k$. Then, for*

$$\tfrac{1}{5}P^k \le M \le P^k, \tag{9.3}$$

*we have*

$$\int_{\mathfrak{M}} T(\alpha)^s e(-M\alpha)\, d\alpha \gg P^{s-k}, \tag{9.4}$$

*where $\mathfrak{M}$ denotes the totality of the major arcs $\mathfrak{M}_{a,q}$.*

**Note.** The reason why we want the result for a range of values of $M$, instead of a single number, will appear later; it will spare us from having to approximate to a somewhat complicated exponential sum on the major arcs.

*Proof.* For $\alpha$ on a particular interval $\mathfrak{M}_{a,q}$, we have (9.2), and the first step is to raise the approximation to the power $s$. We have

$$q^{-1}|S_{a,q}| \ll q^{-1/k}$$

by Lemma 6.4; also

$$I(\beta) \ll \min(P, |\beta|^{-1/k}).$$

The first estimate, $P$, is trivial, and the second comes from writing $I(\beta)$ as

$$\frac{1}{k}|\beta|^{-1/k} \int_0^{P^k|\beta|} e(\pm\eta)\eta^{-1+1/k}\, d\eta$$

and observing that the last integral is bounded. Hence the main term in (9.2) has absolute value

$$\ll q^{-1/k} \min\left(P, |\beta|^{-1/k}\right).$$

The error term $q$ does not exceed this, since

$$q^{1+1/k} < P \quad \text{and} \quad q^{1+1/k} < |\beta|^{-1/k}$$

[1] See [18, Lemmas 8 and 9].

by (9.1). Hence

$$T(\alpha)^s = (q^{-1}S_{a,q}I(\beta))^s + O\left\{q\, q^{-(s-1)/k}\left(\min\left(P, |\beta|^{-1/k}\right)\right)^{s-1}\right\}.$$

The error term here, when integrated with respect to $\beta$ (the range of $\beta$ is immaterial — one can take $(-\infty, \infty)$) becomes

$$\ll q^{1-(s-1)/k}P^{s-1-k}.$$

When this is summed over $a$ (at most $q$ values) and over $q \le P^{1/2}$, it gives a final error term

$$\ll P^{s-1-k}\sum_q q^{2-(s-1)/k} \ll P^{s-k-1},$$

since the series is convergent ($s-1 > 3k$).

The contribution of the main term to the integral in (9.4) is

$$\sum_q \sum_a (q^{-1}S_{a,q})^s e(-Ma/q)\int I^s(\beta)e(-M\beta)\,d\beta,$$

where the conditions of summation and integration are determined by (9.1). We can extend the integration over $\beta$ to $(-\infty, \infty)$, since by the estimate for $|I(\beta)|$ the resulting error is

$$\begin{aligned}&\ll \sum_q q q^{-s/k}\int_{(2kq)^{-1}P^{1-k}}^{\infty} \beta^{-s/k}\,d\beta\\ &\ll \sum_q q^{1-s/k}q^{s/k-1}P^{(k-1)(s/k-1)}\\ &\ll P^{s-k-s/k+3/2} \ll P^{s-k-1}.\end{aligned}$$

By a simple change of variable,

$$I(\beta) = Pk^{-1}\int_0^1 e(\beta P^k\eta)\eta^{-1+1/k}\,d\eta = Pk^{-1}I_1(\beta P^k),$$

say. Putting $\beta = P^{-k}\gamma$, we obtain

$$\int_{-\infty}^{\infty} I^s(\beta)e(-M\beta)\,d\beta = P^{s-k}k^{-s}\int_{-\infty}^{\infty} I_1^s(\gamma)e(-\theta\gamma)\,d\gamma,$$

where $\theta = M/P^k$, so that $\frac{1}{5} \le \theta \le 1$. As in the proof of Theorem 4.1, it follows from Fourier's integral theorem that the integral

$$\int_{-\infty}^{\infty} I_1^s(\gamma)e(-\theta\gamma)\,d\gamma$$

is equal to

$$\int_0^1 \cdots \int_0^1 \{\zeta_1 \cdots \zeta_{s-1}(\theta - \zeta_1 - \cdots - \zeta_{s-1})\}^{-1+1/k} \, d\zeta_1 \cdots \, d\zeta_{s-1},$$

where the integral is taken over $\zeta_1, \ldots, \zeta_{s-1}$ for which $\zeta_1 + \cdots + \zeta_{s-1} < \theta$. Hence

$$\int_{-\infty}^{\infty} I_1^s(\gamma) e(-\theta\gamma) \, d\gamma = \theta^{s/k-1} \frac{\Gamma(1/k)^s}{\Gamma(s/k)},$$

and since $\theta \geq 1/5$, we obtain, on substitution,

$$\int_{-\infty}^{\infty} I^s(\beta) e(-M\beta) \gg P^{s-k}.$$

It suffices now to obtain a positive lower bound for

$$\sum_{q \leq P^{1/2}} \sum_{\substack{a=1 \\ (a,q)=1}}^{q} (q^{-1} S_{a,q})^s e(-Ma/q).$$

This series can be continued to infinity with an error which is bounded by a negative power of $P$; this follows from Lemma 6.4 since $s \geq 4k > 2k+1$. It then becomes $\mathfrak{S}(M)$, and this has a positive lower bound independent of $M$ by Theorem 6.1. Thus Lemma 9.3 is proved. □

We now come to the main idea of the proof. This is: to consider the representations of a large number $N$ in the form

$$N = x_1^k + \cdots + x_{4k}^k + u_1 + u_2 + y^k v, \tag{9.5}$$

where

(i) $1 \leq x_j \leq P$;

(ii) $u_1$ and $u_2$ run through all the different numbers less than $\frac{1}{4}P^k$ that are representable as sums of $\ell$ positive integral $k$th powers;

(iii) $1 \leq y \leq P^{1/2k}$;

(iv) $v$ runs through the different numbers less than $\frac{1}{4}P^{k-1/2}$ that are representable as sums of $\ell$ positive integral $k$th powers.

Thus we shall be representing $N$ as a sum of $4k + 3\ell$ positive integral $k$th powers. In order to prove that a representation exists, we shall have to choose $\ell$ in terms of $k$ later. We shall choose

$$P = [N^{1/k}] + 1;$$

this will ensure that

$$\tfrac{1}{5}P^k < N - u_1 - u_2 - y^k v < P^k. \tag{9.6}$$

It is vital to have a lower bound for the number of $u_1$, $u_2$, $v$, and such a lower bound is provided by the following lemma.

**Lemma 9.4. (Hardy and Littlewood)** *Let $U_\ell(X)$ denote the number of different numbers up to $X$ that are representable as sums of $\ell$ positive integral $k$th powers. Then, provided $X > X_0(k,\ell)$, we have*

$$U_\ell(X) \gg X^{1-\lambda^\ell}, \quad \lambda = 1 - \tfrac{1}{k}. \tag{9.7}$$

*Proof.* The result holds when $\ell = 1$, since then the number is $[X^{1/k}]$, and the number on the right of (9.7) is $X^{1/k}$. The general result is proved by induction on $\ell$. Consider numbers of the form $x^k + z$, where

$$\left(\tfrac{1}{4}X\right)^{1/k} < x < \left(\tfrac{1}{2}X\right)^{1/k}$$

and

$$0 < z < \tfrac{1}{2}X^{1-1/k},$$

and $z$ is expressible as a sum of $\ell - 1$ positive integral $k$th powers. These numbers are all distinct, for if

$$x_1^k + z_1 = x_2^k + z_2, \quad x_1 > x_2,$$

we get

$$x_1^k - x_2^k > kx_2^{k-1} > k(\tfrac{1}{4}X)^{1-1/k} > \tfrac{1}{2}X^{1-1/k},$$

whereas

$$z_2 - z_1 < z_2 < \tfrac{1}{2}X^{1-1/k},$$

a contradiction. The number of possibilities for $x$ is $\gg X^{1/k}$, and the number for $z$ is $U_{\ell-1}\left(\tfrac{1}{2}X^{1-1/k}\right)$, so we have

$$U_\ell(X) \gg X^{1/k} U_{\ell-1}\left(\tfrac{1}{2}X^{1-1/k}\right).$$

If the analogue of (9.7) with $\ell - 1$ in place of $\ell$ holds, we get

$$U_\ell(X) \gg X^{1/k} X^{(1-1/k)(1-\lambda^{\ell-1})} = X^{1-\lambda^\ell},$$

so (9.7) itself holds. This proves the result. □

**Note.** Nothing substantially better than (9.7) is known for general $k$, and any real improvement would be of interest. Better results are known for small $k$ (see the paper of Davenport [19]).

**Corollary.** *Let*

$$R(\alpha) = \sum_{u < \frac{1}{4}P^k} e(\alpha u) \tag{9.8}$$

*where $u$ runs through different numbers that are sums of $\ell$ $k$th powers. Then*

$$\int_0^1 |R(\alpha)|^2 \, d\alpha = R(0) \ll P^{-k(1-\lambda^\ell)} R^2(0). \tag{9.9}$$

*Proof.* The first result is immediate, being valid for any exponential sum $\sum e(\alpha u)$ in which $u$ runs through a set of different integers. The second result follows from the fact that

$$R(0) = U_\ell\left(\tfrac{1}{4}P^k\right) \gg P^{k(1-\lambda^\ell)}.$$

□

**Note.** It is convenient to use $R(0)$ as a means of indicating the number of terms in the exponential sum $R(\alpha)$, in order to avoid introducing new symbols. The trivial estimate for the integral in (9.9) would be $R^2(0)$, and it will be seen that in comparison with this we have saved an amount $k(1 - \lambda^\ell)$ in the exponent of $P$. We shall ultimately choose $\ell$ so that $\lambda^\ell$ (that is, $(1 - 1/k)^\ell$) is about $1/Ck^2$, so that the saving will be about $k - 1/Ck$. It will be necessary to save a further amount, more than $1/Ck$, when $\alpha$ is in the minor arcs $\mathfrak{m}$, and this will be attained by an estimate for the exponential sum corresponding to the last term $y^k v$ in (9.5). The estimate depends on the following very general lemma, the principle of which plays a large part in most of Vinogradov's work.

**Lemma 9.5. (Vinogradov)** *Let $x$ run through a set of $X_0$ distinct integers contained in an interval of length $X$, and let $y$ run through a set of $Y_0$ distinct integers contained in an interval of length $Y$. Suppose $\alpha = a/q + O(q^{-2})$, where $(a, q) = 1$, $q > 1$. Then*

$$\left|\sum_x \sum_y e(\alpha x y)\right|^2 \ll X_0 Y_0 \frac{\log q}{q}(q + X)(q + Y). \tag{9.10}$$

*Proof.* Denoting the sum by $S$, we have

$$|S|^2 \le \left(\sum_x 1\right)\left(\sum_x \left|\sum_y e(\alpha x y)\right|^2\right)$$

$$\leq X_0 \sum_{x=x_1}^{x_1+X} \sum_y \sum_{y'} e(\alpha x(y-y')),$$

where now $x$ runs through *all* integers of the interval containing the original set, whereas $y$, $y'$ are still restricted. Carrying out the summation over $x$, we obtain

$$|S|^2 \ll X_0 \sum_y \sum_{y'} \min(X, \|\alpha(y-y')\|^{-1}).$$

Now $|y-y'| \leq Y$, and each value of $y-y'$ arises from at most $Y_0$ pairs $y$, $y'$. Hence

$$|S|^2 \ll X_0 Y_0 \sum_{\substack{t \\ |t| \leq Y}} \min(X, \|\alpha t\|^{-1}).$$

The rest of the argument is essentially the same as that in the proof of Lemma 3.1. The sum over $t$ is divided into blocks of $q$ consecutive terms, the number of blocks being $\ll Y/q+1$. The sum of the terms in any one block is of the form

$$\sum_{t=t_1+1}^{t_1+q} \min\left(X, \left\|\frac{at}{q} + \tau + O\left(\frac{1}{q}\right)\right\|^{-1}\right)$$
$$\ll X + \sum_{1 \leq u \leq \frac{1}{2}q} \frac{q}{u} \ll X + q \log q.$$

Hence

$$|S|^2 \ll X_0 Y_0 \left(Y/q+1\right)\left(X + q\log q\right),$$

and this implies (9.10). □

**Corollary.** *Let*

$$S(\alpha) = \sum_y \sum_v e(\alpha y^k v), \tag{9.11}$$

*where the conditions of summation are those in (iii) and (iv) earlier. Then, if $\alpha = a/q + O(q^{-2})$, and*

$$P^{1/2} < q \leq 2kP^{k-1},$$

*we have*

$$|S(\alpha)| \ll S(0) P^{-\frac{1}{4k} + \frac{1}{2}(k-\frac{1}{2})\lambda^\ell}. \tag{9.12}$$

*Proof.* The sum is an instance of that of the lemma, with

$$X = \tfrac{1}{4}P^{k-\frac{1}{2}}, \quad X_0 = U_\ell\left(\tfrac{1}{4}P^{k-\frac{1}{2}}\right),$$
$$Y = P^{1/2}, \qquad Y_0 = P^{\frac{1}{2k}}.$$

Hence

$$|S(\alpha)|^2 \ll X_0 P^{\frac{1}{2k}} \frac{\log q}{q}\left(q + \tfrac{1}{4}P^{k-\frac{1}{2}}\right)\left(q + P^{1/2}\right)$$
$$\ll X_0 P^{\frac{1}{2k}} (\log P) P^{k-\frac{1}{2}}.$$

Since

$$S(0) \gg P^{\frac{1}{2k}} X_0,$$

we obtain

$$\left|\frac{S(\alpha)}{S(0)}\right|^2 \ll P^{-\frac{1}{2k}+(k-\frac{1}{2})+\varepsilon} X_0^{-1},$$

and since

$$X_0 \gg P^{(k-\frac{1}{2})(1-\lambda^\ell)}$$

by Lemma 9.4, we obtain (9.12). □

**Note.** It will be seen that (9.12) represents a saving over the trivial estimate of almost $1/4k$ in the exponent of $P$, provided $\lambda^\ell$ is small.

**Lemma 9.6.** *Let $\mathfrak{m}$ denote the minor arcs, that is, the complement of the intervals $\mathfrak{M}_{a,q}$ in (9.1). Then provided $\ell \geq 2k \log 3k$, we have*

$$\int_{\mathfrak{m}} |T(\alpha)|^{4k} |R(\alpha)|^2 |S(\alpha)| \, d\alpha \ll P^{3k} R^2(0) S(0) P^{-\delta}$$

*for some fixed $\delta > 0$.*

*Proof.* For every $\alpha$ there exist $a$, $q$ such that

$$(a, q) = 1, \quad 1 \leq q \leq 2kP^{k-1}, \quad \left|\alpha - \frac{a}{q}\right| < \frac{1}{2kqP^{k-1}},$$

and if $\alpha$ is not in any $\mathfrak{M}_{a,q}$, we must have $q > P^{1/2}$. Note that $|\alpha - a/q| < 1/(2kqP^{k-1})$. By the Corollary to Lemma 9.5,

$$|S(\alpha)| \ll S(0) P^{-\frac{1}{4k}+\frac{1}{2}(k-\frac{1}{2})\lambda^\ell}.$$

By the Corollary to Lemma 9.4,

$$\int_0^1 |R(\alpha)|^2 \, d\alpha \ll R^2(0) P^{-k+k\lambda^\ell}.$$

Using the trivial estimate $|T(\alpha)| \leq P$, we see that the integral in the enunciation is

$$\ll P^{3k} R^2(0) S(0) P^{-\frac{1}{4k} + \frac{3}{2} k \lambda^\ell}.$$

If $\ell \geq 2k \log 3k$, then

$$\log \lambda^\ell = \ell \log \left(1 - \frac{1}{k}\right) < -\frac{\ell}{k} < -2 \log 3k,$$

whence $\lambda^\ell < (3k)^{-2}$ and

$$\frac{3}{2} k \lambda^\ell < \frac{1}{6k}.$$

Hence the result. □

**Theorem 9.1.** $G(k) < 4k + 6k \log 3k + 3$.

*Proof.* Let $r_1(N)$ denote the number of representation of $N$ in the special form (9.5), subject to the conditions given there. Then

$$r_1(N) = \int_0^1 T^{4k}(\alpha) R^2(\alpha) S(\alpha) e(-N\alpha) \, d\alpha.$$

By Lemma 3.1, the contribution of the minor arcs $\mathfrak{m}$ to the integral is of lower order than $P^{3k} R^2(0) S(0)$, provided $\ell \geq 2k \log 3k$.

We write the contribution of the major arcs as

$$\sum_{u_1} \sum_{u_2} \sum_{y} \sum_{v} \int_{\mathfrak{M}} T^{4k}(\alpha) e\left(\alpha(-N + u_1 + u_2 + y^k v)\right) \, d\alpha.$$

The number of terms in the outside sum is $R^2(0) S(0)$, and for any choice of these we have

$$\tfrac{1}{5} P^k < N - u_1 - u_2 - y^k v < P^k,$$

as noted earlier. Thus, by Lemma 9.3, the integral over $\mathfrak{M}$ is $\gg P^{3k}$, since now $s = 4k$. Hence $r_1(N) \gg P^{3k} R^2(0) S(0)$, so that $r_1(N) \to \infty$ as $N \to \infty$. It follows that

$$G(k) \leq 4k + 3\ell \leq 4k + 3(2k \log 3k + 1),$$

giving the result. □

# 10

# The equation $c_1x_1^k + \cdots + c_sx_s^k = 0$ again

We return to this equation, and adapt to it the method of the last section, so as to establish its solubility under a less restrictive condition than that of Theorem 8.1. As in Chapter 8 we shall suppose that $c_1, \ldots, c_s$ are fixed integers, none of them 0, and not all of the same sign if $k$ is even.

Let $s_0 = s_0(k)$ be an integer which has the property that the singular series for any equation

$$c_1x_1^k + \cdots + c_{s_0}x_{s_0}^k = 0$$

is positive. By the work of Chapter 8, we can take

$$s_0 = k^2(2k-1) + 1, \tag{10.1}$$

and as remarked there it is in fact possible to take $s_0 = k^2 + 1$, though we have not proved this.

As in the preceding chapter, it will be necessary to have some knowledge about the singular series $\mathfrak{S}(M)$ of the related equation

$$c_1x_1^k + \cdots + c_{s_0}x_{s_0}^k = M. \tag{10.2}$$

This factorizes as $\prod_p \chi(p)$, and we know that

$$\prod_{p>p_0} \chi(p) \geq \frac{1}{2}$$

for some $p_0$ depending on $k$ and $s$ but not on $M$; this holds if $s_0$ is merely $\geq 2k+1$. Thus to ensure that $\mathfrak{S}(M)$ has a positive lower bound, independent of $M$, for some class of integers $M$, it will suffice if $\chi(p)$ has such a lower bound, for each $p \leq p_0$.

In the work of Chapter 8, we applied (for each $p$) a preliminary transformation to the additive form, depending on the powers of $p$ dividing the coefficients $c_j$. After this we obtained a form

$$d_1 y_1^k + \cdots + d_v y_v^k$$

with coefficients not divisible by $p$, and it was sufficient if the congruence

$$d_1 y_1^k + \cdots + d_v y_v^k \equiv 0 \pmod{p^\gamma}$$

had a solution with $y_1, \ldots, y_v$ not all divisible by $p$. We showed that this condition was satisfied if $s \geq s_0$, with $s_0$ in (10.1).

It follows that for each $p$ there is some $\gamma_1(p)$ (depending also on the coefficients $c_j$) such that $\chi(p)$ has a positive lower bound if[1]

$$M \equiv 0 \pmod{p^{\gamma_1(p)}},$$

for this will ensure that the final congruence which has to be solved is the same as it would be for $M = 0$. Let

$$L = \prod_{p \leq p_0} p^{\gamma_1(p)}.$$

Then if $M \equiv 0 \pmod{L}$ there is a positive lower bound for $\chi(p)$, and hence also for $\mathfrak{S}(M)$, independent of the particular $M$.

Returning to the equation of the title, we take $s = s_0 + 3\ell$, with $\ell$ as in Chapter 9, and divide the coefficients into sets:

$$c_1, \ldots, c_{s_0}; \quad d_1, \ldots, d_\ell; \quad d'_1, \ldots, d'_\ell; \quad e_1, \ldots, e_\ell;$$

subject to the condition that $c_1, \ldots, c_{s_0}$ are not all of the same sign. We shall establish the solubility of the equation

$$c_1 x_1^k + \cdots + c_{s_0} x_{s_0}^k + L^k(u_1 + u_2 + y^k v) = 0, \tag{10.3}$$

subject to

(i) $0 < x_j \leq P/|c_j|^{1/k}$;
(ii) $u_1$ is an integer $< \frac{1}{4}(P/L)^k$, which is representable as $d_1 z_1^k + \cdots + d_\ell z_\ell^k$, and similarly for $u_2$ with accented coefficients;
(iii) $1 < y < (P/L)^{1/2k}$;
(iv) $v$ is an integer $< \frac{1}{4}(P/L)^{k-1/2}$ which is representable as $e_1 t_1^k + \cdots + e_\ell t_\ell^k$.

This will prove the solubility of the original equation when $s = s_0 + 3\ell$, and *a fortiori* when $s \geq s_0 + 3\ell$.

The definition of major and minor arcs is the same as in Chapter 9, except that we replace $2kqP^{k-1}$ by $2kqcP^{k-1}$, where $c = \max |c_j|$; this

[1] We can take $\gamma_1(p) = \gamma$ if $p$ does not divide any of the $c_j$.

is to ensure that the conditions of van der Corput's lemma are satisfied in the proof of the analogue of Lemma 9.2. In place of Lemma 9.3 we obtain that

$$\int_{\mathfrak{M}} T_1(\alpha)\cdots T_{s_0}(\alpha)e(-M\alpha)d\alpha \gg P^{s_0-k} \tag{10.4}$$

provided $0 < M \le P^k$ and $M \equiv 0 \pmod{L}$. The proof of this differs only in respect of the singular integral, which transforms into a multiple of

$$\int_0^1 \cdots \int_0^1 \{\zeta_1 \cdots \zeta_{s_0-1}(\pm\zeta_1 \pm \cdots \pm \zeta_{s_0-1} - \theta)\}^{1-1/k} d\zeta_1 \cdots d\zeta_{s_0-1},$$

integrated over the range $0 < \pm\zeta_1 \pm \cdots \pm \zeta_{s_0-1} - \theta < 1$, where the signs are those of $c_1, \ldots, c_{s_0-1}$ and we have assumed that $c_{s_0}$ is negative. We can suppose that two at least of the signs $\pm$ are $+$. Then, for $0 \le \theta \le 1$, the region of integration contains some small cube of size independent of $\theta$, and it follows that the above integral has a positive lower bound.

We have already seen that the singular series occurring in the analogue of Lemma 9.3 has a positive lower bound for $M \equiv 0 \pmod{L}$, and hence we obtain (10.4).

Lemma 9.4 still applies, with slight changes, to give lower bounds for the number of integers of the form $u$ or $v$. We consider the numbers $u$ representable as

$$d_1z_1^k + z,$$

where $z$ is representable as $d_2z_2^k + \cdots + d_\ell z_\ell^k$, and, if the variables $z_1$ and $z$ are restricted to suitable ranges, the numbers are all distinct. The ranges depend of course on $d_1, \ldots, d_\ell$. We get the same lower bound for $U_\ell(X)$ as before, apart from a constant depending on $d_1, \ldots, d_\ell$.

We now need two exponential sums $R_1(\alpha)$, $R_2(\alpha)$ corresponding to the two different sets of numbers $u_1, u_2$; but since

$$\int_0^1 |R_1(\alpha)R_2(\alpha)|d\alpha \le \left\{\int_0^1 |R_1(\alpha)|^2 d\alpha \int_0^1 |R_2(\alpha)|^2 d\alpha\right\}^{1/2},$$

we get the same saving as in the Corollary to Lemma 9.4.

The Corollary to Lemma 9.5 is essentially unchanged, and the proof is completed as before. Thus we obtain the following result.

**Theorem 10.1.** *Let* $c_1, \ldots, c_s$ *be integers, none of them* 0 *and not all of the same sign if* $k$ *is even. Then, if*

$$s \ge s_0 + 3(2k\log 3k + 1), \tag{10.5}$$

*where* $s_0$ *has the value given in (10.1), the equation*

$$c_1 x_1^k + \cdots + c_s x_s^k = 0$$

*has infinitely many solutions in integers* $x_1, \ldots, x_s$, *none of them* $0$.

# 11

# General homogeneous equations: Birch's theorem

We now pass to homogeneous equations in general, that is, equations which are not necessarily of the additive type considered so far. Let $f(x_1, \ldots, x_n)$ be a homogeneous polynomial, which we call a *form*, of degree $k$ with integral coefficients. We are interested in the solubility of

$$f(x_1, \ldots, x_n) = 0$$

in integers $x_1, \ldots, x_n$ (not all 0). Owing to the homogeneity, we can allow the coefficients and variables to be rational instead of integral without changing the question. An obvious necessary condition is that the equation must be soluble in real numbers, not all 0.

When $k = 2$, so that $f$ is a quadratic form, the general form can be expressed as an additive form by the process of 'completing the square'. It is known from the classical theory of quadratic forms that the the congruence conditions which are obviously necessary for solubility, namely the conditions that $f \equiv 0 \pmod{p^\nu}$ shall be soluble for every prime power $p^\nu$ with not all the variables divisible by $p$, together with the condition that the form $f$ shall be indefinite, are also sufficient. The congruence conditions are significant only for a finite set of primes $p$, the set depending on the coefficients, and it is necessary that the congruences shall have a solution with $x_1, \ldots, x_n$ not all divisible by $p$; here also $\nu$ is determined by the coefficients. These congruence conditions are always satisfied if $n \geq 5$, accordingly we have the well known theorem of Meyer (1883) that an indefinite quadratic form in five or more variables always represents zero non-trivially.

For $k > 2$, the general form is of far greater generality than the additive form. For instance, if $k = 3$, there are $\frac{1}{6}n(n+1)(n+2)$ independent coefficients in the general form, and since a linear transformation on $n$

variables has only $n^2$ coefficients, it is plain that little simplification is possible.

An important question in connection with general forms is that of degeneration. A form in $n$ variables is said to *degenerate* if there is a non-singular linear transformation from $x_1, \ldots, x_n$ to $y_1, \ldots, y_n$ such that the transformed form is one in $y_1, \ldots, y_{n-1}$ only; that is, the coefficients of all terms containing $y_n$ vanish. Degeneration is an absolute property, in the sense that if a form does not degenerate by substitutions with coefficients in a particular field (e.g. the rational field), this will remain true if the field is extended. For in the above notation, we must have $\partial f/\partial y_n = 0$ identically, and if $C_j$ is the coefficient of $y_n$ in $x_j$, this is equivalent to

$$C_1 \frac{\partial f}{\partial x_1} + \cdots + C_n \frac{\partial f}{\partial x_n} = 0$$

identically. This identity represents a finite number of linear relations in the coefficients $C_1, \ldots, C_n$, and if these can be satisfied at all, they can be satisfied in the original field.

However, from the point of view of the present problem – that of representing zero – we can always suppose $f$ non-degenerate; for if $f$ degenerates as above, there is a solution with $x_1, \ldots, x_n$ not all 0 corresponding to the solution $y_1 = \cdots = y_{n-1} = 0$, $y_n = 1$.

The first substantial progress towards solving general homogeneous equations was made by R. Brauer [9]. He showed that, for equations with the coefficients and variables in any field, the solubility of a homogeneous equation of degree $k$, or of a system of homogeneous equations of degree $\leq k$, can be deduced from the solubility of every additive equation of degree $\leq k$. But the number of variables in the original equation has to be taken to be enormously large in order to get a reasonable number of variables in the final equations. In itself, the theorem cannot be applied with success in the rational field, because among the additive equations there will be some of even degree, and these will certainly not be soluble if the coefficients are all of the same sign. Brauer's theorem applies, however, in the $p$-adic number field, and establishes the existence of a function $n_1(k, h)$ such that any system of $h$ homogeneous equations of degrees $\leq k$ in $n$ variables, with $p$-adic coefficients, is soluble in $p$-adic numbers if $n \geq n_1(k, h)$.

The simplest problem in this field is that of a single cubic equation. Professor Lewis [56] was the first to prove that there exists an absolute constant $n_0$ such that any cubic equation $f = 0$ in $n$ variables, with rational coefficients, is soluble in rational numbers if $n \geq n_0$. He deduced this from Brauer's result by supplementing the latter by arguments based on algebraic number theory.

Shortly afterwards, Birch [5] proved a more general theorem, namely that any system of homogeneous equations, each of odd degree, is soluble provided $n$ exceeds a certain function of the degrees. This theorem forms the subject of the present chapter.

Before proving Birch's theorem in all its generality, I propose to explain the method in relation to the simplest case: that of a single cubic equation. It will be found that we need only to solve certain systems of linear equations in order to deduce the solubility of the general cubic equation from that of additive cubic equations. The solubility of the latter is assured by Theorem 8.1, provided the number of variables is at least $k^2(2k-1)+1 = 46$. (Actually 8 variables would suffice, but this needs special arguments.) As with Brauer's method, of which Birch's is an ingenious modification, there is a considerable wastage of variables in getting from the general form to the additive form.

Let$f(x_1, \ldots, x_n)$ be a cubic form. We say that $f$ represents $g(y_1, \ldots, y_m)$, where $m \leq n$, if there exist $n$ linear forms in $y_1, \ldots, y_m$, with rank $m$ such that when $x_1, \ldots, x_n$ are replaced by these linear forms, we get

$$f(x_1, \ldots, x_n) = g(y_1, \ldots, y_m)$$

identically in $y_1, \ldots, y_m$. The essential idea of the proof is to show that $f$ represents a form of the type

$$a_0 t_0^3 + g_1(t_1, \ldots, t_m). \tag{11.1}$$

provided $n$ exceeds a certain number depending only on $m$.

Write

$$f(\mathbf{x}) = f(x_1, \ldots, x_n) = \sum_i \sum_j \sum_k c_{ijk} x_i x_j x_k,$$

where the sums go from 1 to $n$ and where $c_{ijk}$ is a symmetrical function of the three subscripts. Define the trilinear function $T(\mathbf{x} \mid \mathbf{y} \mid \mathbf{z})$ of three points $\mathbf{x}$, $\mathbf{y}$, $\mathbf{z}$, by

$$T(\mathbf{x} \mid \mathbf{y} \mid \mathbf{z}) = \sum_i \sum_j \sum_k c_{ijk} x_i y_j z_k.$$

We choose any $m$ linearly independent points $\mathbf{y}^{(1)}, \ldots, \mathbf{y}^{(m)}$, with integral (or rational) coordinates, and consider the equations

$$T(\mathbf{y} \mid \mathbf{y}^{(p)} \mid \mathbf{y}^{(q)}) = 0, \quad 1 \leq p \leq q \leq m$$

in an unknown point $\mathbf{y}$. These are $\frac{1}{2}m(m+1)$ linear equations in the $n$ coordinates of $\mathbf{y}$, so if $n > \frac{1}{2}m(m+1)$ there is a point $\mathbf{y}$ other than the origin which satisfies them. Calling such a point $\mathbf{y}^{(0)}$, we have

$$T(\mathbf{y}^{(0)} \,|\, \mathbf{y}^{(p)} \,|\, \mathbf{y}^{(q)}) = 0$$

for $1 \le p \le q \le m$.

If the points $\mathbf{y}^{(0)}$, $\mathbf{y}^{(1)}$, ..., $\mathbf{y}^{(m)}$ were linearly dependent, then (since the last $m$ are linearly independent) we should have

$$\mathbf{y}^{(0)} = c_1\mathbf{y}^{(1)} + \cdots + c_m\mathbf{y}^{(m)}$$

with rational numbers $c_1, \ldots, c_m$. But then

$$T(\mathbf{y}^{(0)} \,|\, \mathbf{y}^{(0)} \,|\, \mathbf{y}^{(0)}) = \sum_{p=1}^{m}\sum_{q=1}^{m} c_p c_q T(\mathbf{y}^{(0)} \,|\, \mathbf{y}^{(p)} \,|\, \mathbf{y}^{(q)}) = 0,$$

whence $f(\mathbf{y}^{(0)}) = 0$, giving the desired solution.

So we can suppose that the $m+1$ points are linearly independent. The transformation

$$x_j = t_0 y_j^{(0)} + t_1 y_j^{(1)} + \cdots + t_m y_j^{(m)}, \quad (1 \le j \le n)$$

has rank $m+1$, and expresses $f$ as a form in $t_0, t_1, \ldots, t_m$. The coefficient of $t_0 t_p t_q$ in this form is $T(\mathbf{y}^{(0)} \,|\, \mathbf{y}^{(p)} \,|\, \mathbf{y}^{(q)})$, and so is 0 for $1 \le p \le q \le m$. Hence the new form is of the shape

$$a_0 t_0^3 + t_0^2(b_1 t_1 + \cdots + b_m t_m) + g_1(t_1, \ldots, t_m),$$

there being no terms of degree 1 in $t_0$. If $b_1, \ldots, b_m$ are all 0, we have a form of the desired type (11.1). If not, say $b_m \ne 0$, we put

$$t_m = -\frac{1}{b_m}(b_1 t_1 + \cdots + b_{m-1} t_{m-1})$$

and obtain a form of the type (11.1) but with $m-1$ instead of $m$. Hence $f$ represents a form of the type (11.1) provided only that $n > \frac{1}{2}m(m+1)$. We can suppose in (11.1) that $a_0 \ne 0$, for if $a_0 = 0$ there is a solution in $x_1, \ldots, x_n$ corresponding to $t_0 = 1$, $t_1 = \cdots = t_m = 0$.

We see that we can have $m$ as large as we please by taking $n$ sufficiently large. Also we can repeat the process on the new form $g(t_1, \ldots, t_m)$, which can be assumed not to represent zero. It follows that there is a function $n_0(s)$ such that, if $n \ge n_0(s)$, the form $f$ represents a form of the type

$$b_1 u_1^3 + \cdots + b_s u_s^3.$$

By Theorem 8.1, the corresponding homogeneous equation has a non-trivial solution if $s \geq s_0$ ($= 46$, say), and hence the original equation is soluble if $n \geq n_0(s_0)$.

We now prove Birch's general theorem:

**Theorem 11.1.** *Let $h$, $m$ be positive integers, and let $r_1, \ldots, r_h$ be any odd positive integers. Then there exists a number*

$$\Psi(r_1, \ldots, r_h; m)$$

*with the following property. Let $f_{r_1}(\mathbf{x}), \ldots, f_{r_h}(\mathbf{x})$ be forms of degrees $r_1, \ldots, r_h$ respectively in $\mathbf{x} = (x_1, \ldots, x_n)$, with rational coefficients. Then, provided $n \geq \Psi(r_1, \ldots, r_h; m)$, there is an $m$-dimensional rational vector space, all points in which satisfy*

$$f_{r_1}(\mathbf{x}) = 0, \ldots, f_{r_h}(\mathbf{x}) = 0.$$

Note that we assert more than the existence of an infinity of solutions; this is for convenience in the inductive proof.

**Lemma 11.1.** *There exists a number $\Phi(r, m)$, defined for positive integers $m$, $r$ with $r$ odd, such that if $s \geq \Phi(r, m)$, any equation of the form*

$$c_1 x_1^r + \cdots + c_s x_s^r = 0, \quad (c_j \text{ rational})$$

*is satisfied by all points of some rational linear space of dimension $m$.*

*Proof.* The result is a simple deduction from Theorem 8.1 or Theorem 10.1. By either of these theorems there exists $t = t(r)$ such that the equation

$$c_1 y_1^r + \cdots + c_t y_t^r = 0$$

has a non-zero integral solution. Note that since $r$ is odd, there is no condition on the signs of the coefficients; also if any of them is 0 there is an obvious solution. Similarly for the equation

$$c_{t+1} y_{t+1}^r + \cdots + c_{2t} y_{2t}^r = 0,$$

and so on. Hence, if $s \geq mt$, the point

$$(u_1 y_1, \ldots, u_1 y_t, u_2 y_{t+1}, \ldots, u_2 y_{2t}, \ldots, u_m y_{mt}, 0, \ldots, 0)$$

formed out of these solutions, satisfies the equation for all values of $u_1, \ldots, u_m$. As these vary, the point describes an $m$-dimensional linear space. Hence the result, with $\Psi(r, m) = mt$. □

*Proof of Theorem 11.1.* Let $R = \max r_j$, so that $R$ is an odd positive integer. The result certainly holds when $R = 1$, and we prove it by induction on $R$ (limited to odd values). We shall first prove that if the result holds for systems of equations with $\max r_j \leq R-2$, then it holds for a single equation of degree $R$. Once this has been proved, it will be easy to extend the result to a system of equations of degree $\leq R$, and thereby complete the inductive proof.

For a form $f(x_1, \ldots, x_n)$ of degree $R$, we define a multilinear function of $R$ points $\mathbf{x}^{(1)}, \ldots, \mathbf{x}^{(R)}$ by

$$\begin{aligned} f(x_1, \ldots, x_n) &= \sum c_{i_1,\ldots,i_R} x_{i_1} \ldots x_{i_R}, \\ M(\mathbf{x}^{(1)} \mid \cdots \mid \mathbf{x}^{(R)}) &= \sum c_{i_1,\ldots,i_R} x_{i_1}^{(1)} \ldots x_{i_R}^{(R)}. \end{aligned}$$

We begin by selecting $h$ linearly independent points $\mathbf{y}^{(1)}, \ldots, \mathbf{y}^{(h)}$. Let $\mathcal{Y}$ be the $h$-dimensional linear space generated by them. Consider the equations

$$M(\underbrace{\mathbf{y} \mid \cdots \mid \mathbf{y}}_{\rho} \mid \mathbf{y}^{(p_1)} \mid \cdots \mid \mathbf{y}^{(p_{R-\rho})}) = 0,$$

where $\rho$ takes all odd values from 1 to $R-2$, and $p_1, \ldots, p_{R-\rho}$ take all values from 1 to $h$. The total number of equations is less than $Rh^R$. Each of them is of odd degree $\leq R-2$ in $\mathbf{y}$. Hence, by the inductive hypothesis (for $m = 1$) these equations have a non-zero solution in $\mathbf{y}$ provided $n \geq n_0(R, h)$. Denote such a solution by $\mathbf{y}^{(0)}$. We now have

$$M(\underbrace{\mathbf{y}^{(0)} \mid \cdots \mid \mathbf{y}^{(0)}}_{\rho} \mid \underbrace{\mathbf{y} \mid \cdots \mid \mathbf{y}}_{R-\rho}) = 0$$

for all $y$ in $\mathcal{Y}$ and all odd $\rho \leq R-2$.

An arbitrary point $\mathbf{y}$ in $\mathcal{Y}$ is of the form

$$u_1 \mathbf{y}^{(1)} + \cdots + u_h \mathbf{y}^{(h)}.$$

Now consider the equations

$$M(\underbrace{\mathbf{y}^{(0)} \mid \cdots \mid \mathbf{y}^{(0)}}_{R-\sigma} \mid \underbrace{\mathbf{y} \mid \cdots \mid \mathbf{y}}_{\sigma}) = 0,$$

where $\sigma$ takes all odd values from 1 to $R-2$, and $\mathbf{y}$ is any point in $\mathcal{Y}$. These are equations of odd degree $\leq R-2$ in $u_1, \ldots, u_h$, and their number is $< R$. By the inductive hypothesis again, given any $\ell$, these equations will be satisfied at all points of some rational linear subspace of $\mathcal{Y}$ of dimension $\ell+1$, provided $h \geq h_0(R, \ell)$. Denote this linear subspace by $\mathcal{Y}_1$. We can suppose, without loss of generality, that $\mathcal{Y}_1$ is generated by $\mathbf{y}^{(1)}, \ldots, \mathbf{y}^{(\ell+1)}$. We now have

$$M(\underbrace{\mathbf{y}^{(0)} \mid \cdots \mid \mathbf{y}^{(0)}}_{\tau} \mid \underbrace{\mathbf{y} \mid \cdots \mid \mathbf{y}}_{R-\tau}) = 0$$

for all $\tau = 1, \ldots, R-1$ and for all $\mathbf{y}$ in the $(\ell+1)$-dimensional space $\mathcal{Y}_1$. To ensure this we need only suppose that $n \geq n_0(R, h_0(R, \ell)) = n_1(R, \ell)$.

If $\mathbf{y}^{(0)}$ is in the subspace $\mathcal{Y}_1$, generated by $\mathbf{y}^{(1)}, \ldots, \mathbf{y}^{(\ell+1)}$, we can obtain an $\ell$-dimensional subspace $\mathcal{Y}_2$ of $\mathcal{Y}_1$ which does not contain $\mathbf{y}^{(0)}$ by omitting one of the generating points $\mathbf{y}^{(1)}, \ldots, \mathbf{y}^{(\ell+1)}$, say the last. Now the points $\mathbf{y}^{(0)}, \mathbf{y}^{(1)}, \ldots, \mathbf{y}^{(\ell)}$ are linearly independent. The linear transformation

$$x_j = t_0 y_j^{(0)} + \cdots + t_\ell y_j^{(\ell)}, \quad (1 \leq j \leq n)$$

has rank $\ell + 1$ and gives $f(\mathbf{x}) = g(t_0, \ldots, t_\ell)$, say. The coefficient of $t_0^\tau t_{p_1} \cdots t_{p_{R-\tau}}$ in $g$ (where each $p_j$ goes from 1 to $\ell$) is

$$M(\mathbf{y}^{(0)} \mid \cdots \mid \mathbf{y}^{(0)} \mid \mathbf{y}^{(p_1)} \mid \cdots \mid \mathbf{y}^{(p_{R-\tau})}) = 0,$$

and this holds for $\tau = 1, \ldots, R-1$ and any choice of $p_1, \ldots, p_{R-\tau}$. Hence

$$g(t_0, \ldots, t_\ell) = a_0 t_0^R + g_1(t_1, \ldots, t_\ell).$$

Hence $f$ represents a form of the above type, for any $\ell$, provided $n \geq n_1(R, \ell)$. Repetition of the argument proves that $f$ represents a form of the type

$$\underbrace{a_0 t_0^R + b_0 u_0^R + \cdots}_{s}$$

provided $n \geq n_2(R, s)$. By Lemma 11.1, the solutions of the homogeneous equation obtained by equating the latter form to 0 include a linear space of dimension $m$ provided $s \geq \Phi(r, m)$. Hence the solutions of the original equation include a linear space of dimension $m$ provided $n \geq n_2(R, \Phi(R, m))$. This proves the desired result for a single equation of degree $R$.

Now suppose there are $h$ equations $f_{r_1} = 0, \ldots, f_{r_h} = 0$, where each $r_j \leq R$. We prove the result by induction on $h$, the case $h = 1$ being what we have just proved. Given $m_1$ there exists a rational linear space of dimension $m_1$ on which $f_{r_h} = 0$, provided $n \geq \Psi(r_h; m_1)$, by the case $h = 1$ of the theorem. We can represent the points of this linear space as

$$v_1 \mathbf{x}^{(1)} + \cdots + v_{m_1} \mathbf{x}^{(m_1)}.$$

For these points, the forms $f_{r_1}, \ldots, f_{r_{h-1}}$ become forms in $v_1, \ldots, v_{m_1}$. By the case $h-1$, there is a linear space of dimension $m$ on which they all vanish, provided $m_1 \geq \Psi(r_1, \ldots, r_{h-1}; m)$. Hence all the forms vanish on this space, and the case $h$ of the theorem holds, with

$$\Psi(r_1, \ldots, r_h; m) = \Psi(r_h; \Psi(r_1, \ldots, r_{h-1}; m)).$$

This completes the proof. □

**Corollary.** *Theorem 11.1 continues to hold if the forms have coefficients in an algebraic number field $K$ and $x_1, \ldots, x_n$ have values in $K$, but with $\Psi$ depending now on $K$.*

When we express each coefficient and each variable linearly in terms of a basis of $K$ (say $\{1, \theta, \ldots, \theta^{\nu-1}\}$, where $\nu$ is the degree of $K$ and $\theta$ a generating element), each equation is equivalent to $\nu$ equations of the same degree with rational coefficients and variables.

Our next aim will be to prove a more precise result concerning the number of variables which will suffice to make one homogeneous cubic equation soluble. In the course of the work we shall need some results from the geometry of numbers, and so we begin with an account of certain aspects of this subject.

# 12

# The geometry of numbers

We now prove some of the basic results of the geometry of numbers, limiting ourselves to those which help one to handle linear inequalities. A fuller exposition (such as that given by Cassels [12]) would be both more general and more precise: more general in that ordinary distances are replaced by distances in a metric, and more precise in that attention is paid to constants depending on $n$ (the number of dimensions). Such constants are of no importance for the purpose we have in mind.

A lattice $\Lambda$, in $n$-dimensional space, is the set of all (real) points

$$\mathbf{x} = (x_1, \ldots, x_n)$$

given by $n$ linear forms in $n$ variables $u_1, \ldots, u_n$ which take integral values:

$$\begin{aligned} x_1 &= \lambda_{11}u_1 + \cdots + \lambda_{1n}u_n, \\ &\vdots \\ x_n &= \lambda_{n1}u_1 + \cdots + \lambda_{nn}u_n, \end{aligned}$$

or $\mathbf{x} = \Lambda\mathbf{u}$ in matrix notation. The coefficients $\lambda_{ij}$ are real numbers with $\det \lambda_{ij} \neq 0$. A linear integral substitution of determinant 1 (unimodular substitution) applied to the variables $u_1, \ldots, u_n$ does not change the lattice. The points of the lattice can also be represented by

$$\mathbf{x} = u_1\mathbf{x}^{(1)} + \cdots + u_n\mathbf{x}^{(n)},$$

where $\mathbf{x}^{(j)} = (\lambda_{1j}, \ldots, \lambda_{nj})$ for $j = 1, \ldots, n$. The points $\mathbf{x}^{(1)}, \ldots, \mathbf{x}^{(n)}$ constitute a *basis* for the lattice, and the change of variable just mentioned corresponds to a change of basis.

We define the *determinant* $d(\Lambda)$ of $\Lambda$ by

$$d(\Lambda) = |\det \lambda_{ij}|;$$

this is a positive number, unaffected by unimodular changes of variable. The density of the lattice (in an obvious sense) is $1/d(\Lambda)$. The determinant of the coordinates of any set of $n$ lattice points is an integral multiple of $d(\Lambda)$. We shall usually suppose $d(\Lambda) = 1$ for convenience.

An *affine transformation* of space is a (homogeneous) linear transformation, with real coefficients and determinant $\neq 0$, from $x_1, \ldots, x_n$ to $y_1, \ldots, y_n$. This maps the lattice $\Lambda$ in $x$-space into a lattice $\mathsf{M}$ in $y$-space, and if the affine transformation has determinant 1 then $d(\Lambda) = d(\mathsf{M})$.

**Lemma 12.1.** *Any ellipsoid with centre at the origin $O$ and volume $>$ $2^n$ contains a point, other than $O$, of every lattice of determinant* 1.

*Proof.* It suffices to prove the result for a sphere, since an ellipsoid can be transformed into a sphere by an affine transformation of space of determinant 1. Let $\rho$ be the radius of the sphere. Since $d(\Lambda) = 1$, the number of points of $\Lambda$ in a cube of large side $X$ is asymptotic to $X^n$ as $X \to \infty$. If we place a sphere of radius $\frac{1}{2}\rho$ (and therefore of volume $V > 1$) with its centre at each of these lattice points, their total volume is asymptotically $VX^n$. They are all contained in a cube of side $X+\rho$. For large $X$ we have $VX^n > (X+\rho)^n$, so two of the spheres must overlap. Thus there are two distinct points of $\Lambda$ with distance apart less than $\rho$, and so there is a point of $\Lambda$, other than $O$, within a distance $\rho$ of $O$. □

**Note.** This result, with the ellipsoid replaced by any convex body which has central symmetry about $O$, is Minkowski's first fundamental theorem. The proof is essentially the same.

The *successive minima* of a lattice $\Lambda$ are defined as follows. Let $R_1$ be the least distance of any point of $\Lambda$, other than $O$, from $O$, and let $\mathbf{x}^{(1)}$ be a point of $\Lambda$ at this distance. Denoting by $|\mathbf{x}|$ the distance of a point $\mathbf{x}$ from $O$, we have $|\mathbf{x}^{(1)}| = R_1$. Let $R_2$ be the least distance from $O$ of any point which is not on the line $\langle O, \mathbf{x}^{(1)} \rangle$, and let $\mathbf{x}^{(2)}$ be such a point with $|\mathbf{x}^{(2)}| = R_2$. Let $R_3$ be the least distance from $O$ of any point of $\Lambda$ which is not in the plane $\langle O, \mathbf{x}^{(1)}, \mathbf{x}^{(2)} \rangle$, and so on. We obtain numbers $R_1, \ldots, R_n$ and linearly independent points $\mathbf{x}^{(1)}, \ldots, \mathbf{x}^{(n)}$, such that

$$0 < R_1 \leq R_2 \leq \cdots \leq R_n, \qquad |\mathbf{x}^{(\nu)}| = R_\nu.$$

The points $\mathbf{x}^{(1)}, \ldots, \mathbf{x}^{(n)}$ can possibly be chosen in more than one way, but it is easily seen that this does not affect the uniqueness of the numbers $R_1, \ldots, R_n$. These numbers are the successive minima of $\Lambda$, and

$\mathbf{x}^{(1)}, \ldots, \mathbf{x}^{(n)}$ is a set of minimal points of $\Lambda$. These points do not necessarily constitute a basis, though this happens to be the case when $n = 2$.

**Lemma 12.2.** *If $d(\Lambda) = 1$, we have*

$$1 \leq R_1 R_2 \cdots R_n \leq 2^n / J_n, \tag{12.1}$$

*where $J_n$ denotes the volume of a sphere of radius 1 in $n$ dimensions.*

*Proof.* We can rotate the $n$-dimensional space about $O$ until

$$\begin{aligned}
\mathbf{x}^{(1)} &= (x_1^{(1)}, 0, 0, \ldots, 0), \\
\mathbf{x}^{(2)} &= (x_1^{(2)}, x_2^{(2)}, 0, \ldots, 0), \\
&\vdots \\
\mathbf{x}^{(n)} &= (x_1^{(n)}, x_2^{(n)}, x_3^{(n)}, \ldots, x_n^{(n)}).
\end{aligned}$$

Since the determinant of these $n$ points is an integral multiple of $d(\Lambda) = 1$ and they are linearly independent, we have

$$|x_1^{(1)} x_2^{(2)} \cdots x_n^{(n)}| \geq 1.$$

Since $R_\nu = |\mathbf{x}^{(\nu)}| \geq |x_\nu^{(\nu)}|$, we obtain $R_1 R_2 \cdots R_n \geq 1$.

To obtain an upper bound for $R_1 R_2 \cdots R_n$, we consider the ellipsoid

$$\frac{x_1^2}{R_1^2} + \cdots + \frac{x_n^2}{R_n^2} < 1.$$

This contains no point of $\Lambda$ other than $O$. For suppose a point $\mathbf{x}$ of $\Lambda$ is linearly dependent on $\mathbf{x}^{(1)}, \ldots, \mathbf{x}^{(\nu)}$ but not on $\mathbf{x}^{(1)}, \ldots, \mathbf{x}^{(\nu-1)}$, where $1 \leq \nu \leq n$. Then $|\mathbf{x}| \geq R_\nu$ by the definition of $R_\nu$. Also $x_{\nu+1} = 0, \ldots, x_n = 0$; whence

$$\frac{x_1^2}{R_1^2} + \cdots + \frac{x_n^2}{R_n^2} \geq \frac{x_1^2 + \cdots + x_\nu^2}{R_\nu^2} \geq 1.$$

Thus $\mathbf{x}$ is not in the ellipsoid. It follows from Lemma 12.1 that the volume of the ellipsoid is $\leq 2^n$. This volume is $(R_1 R_2 \cdots R_n) J_n$, and the desired inequality follows. □

**Note.** Lemma 12.2, and the definitions concerning successive minima, again extend to an arbitrary convex body with center $O$. In this more general form, Lemma 12.2 is Minkowski's second fundamental theorem; but its proof is then considerably more difficult. Note that Lemma 12.2 implies $J_n R_1^n \leq 2^n$, which is the result of Lemma 12.1.

We introduce temporarily the notation $A \asymp B$ to mean both $A \ll B$ and $A \gg B$; in other words, to indicate that $A/B$ is bounded both above and below by numbers depending only on $n$.

**Lemma 12.3.** *After an appropriate rotation of space, any lattice of determinant* 1 *has a basis* $\mathbf{X}^{(1)}, \ldots, \mathbf{X}^{(n)}$ *of the form*

$$\mathbf{X}^{(1)} = (X_1^{(1)}, 0, 0, \ldots, 0), \quad \mathbf{X}^{(2)} = (X_1^{(2)}, X_2^{(2)}, 0, \ldots, 0), \quad \ldots$$

*where*

$$|\mathbf{X}^{(\nu)}| \asymp R_\nu \qquad \text{and} \qquad |X_\nu^{(\nu)}| \asymp R_\nu, \tag{12.2}$$

*for* $\nu = 1, \ldots, n$.

*Proof.* We obtain the basis points by a process of adaptation from the minimal points $\mathbf{x}^{(1)}, \ldots, \mathbf{x}_n^{(n)}$ in the proof of Lemma 12.2. We take $\mathbf{X}^{(1)}$ to be $\mathbf{x}^{(1)}$. We take $\mathbf{X}^{(2)}$ to be a point of $\Lambda$ in the plane $\langle O, \mathbf{x}^{(1)}, \mathbf{x}^{(2)} \rangle$ which, together with $\mathbf{X}^{(1)}$, generates integrally all points of $\Lambda$ in that plane; the existence of such a point is geometrically intuitive. It is arbitrary to the extent of any added multiple of $\mathbf{x}^{(1)}$. Since $\mathbf{x}^{(1)}, \mathbf{x}^{(2)}$ generate rationally (though perhaps not integrally) all points of $\Lambda$ in the plane $\langle O, \mathbf{x}^{(1)}, \mathbf{x}^{(2)} \rangle$, we have

$$N\mathbf{X}^{(2)} = u_1 \mathbf{x}^{(1)} + u_2 \mathbf{x}^{(2)}$$

for certain integers $N > 0$, $u_1, u_2$. Since $\mathbf{x}^{(2)}$ is an integral linear combination of $\mathbf{X}^{(2)}$ and $\mathbf{X}^{(1)} = \mathbf{x}^{(1)}$, we must have $u_2 = 1$. By adding to $\mathbf{X}^{(2)}$ a suitable integral multiple of $\mathbf{x}^{(1)}$, we can suppose that $|u_1| \le \frac{1}{2}N$. Then

$$|\mathbf{X}^{(2)}| \le \left| \frac{u_1}{N} \right| |\mathbf{x}^{(1)}| + \frac{1}{N} |\mathbf{x}^{(2)}| \le \frac{1}{2} R_1 + R_2 \le \frac{3}{2} R_2.$$

Next we take $\mathbf{X}^{(3)}$ to be a point of $\Lambda$ in the space $\langle O, \mathbf{x}^{(1)}, \mathbf{x}^{(2)}, \mathbf{x}^{(3)} \rangle$ which, together with $\mathbf{X}^{(1)}$, $\mathbf{X}^{(2)}$, generates integrally all points of $\Lambda$ in that space; the choice of $\mathbf{X}^{(3)}$ is arbitrary to the extent of added multiples of $\mathbf{X}^{(1)}, \mathbf{X}^{(2)}$, and *a fortiori* of added multiples of $\mathbf{x}^{(1)}, \mathbf{x}^{(2)}$. For the same reason as before, we have

$$N\mathbf{X}^{(3)} = u_1 \mathbf{x}^{(1)} + u_2 \mathbf{x}^{(2)} + u_3 \mathbf{x}^{(3)}$$

for certain integers $N > 0$, $u_1, u_2, u_3$. This time we cannot conclude that $u_3 = 1$, but we can conclude that $u_3$ divides $N$, for $\mathbf{x}^{(3)}$ is expressible as

$$\mathbf{x}^{(3)} = v_1 \mathbf{X}^{(1)} + v_2 \mathbf{X}^{(2)} + v_3 \mathbf{X}^{(3)},$$

and we must have $N = u_3 v_3$. As before, we can ensure that $|u_1| \le \frac{1}{2}N$ and $|u_2| \le \frac{1}{2}N$. Hence

$$|\mathbf{X}^{(3)}| \le \frac{1}{2}|\mathbf{x}^{(1)}| + \frac{1}{2}|\mathbf{x}^{(2)}| + |\mathbf{x}^{(3)}| \le \frac{1}{2}R_1 + \frac{1}{2}R_2 + R_3 \le 2R_3.$$

Continuing in this way, we get an integral basis for $\Lambda$ satisfying

$$|\mathbf{X}^{(\nu)}| \le \frac{\nu+1}{2} R_\nu \ll R_\nu, \quad (1 \le \nu \le n).$$

Since $\mathbf{X}^{(\nu)}$ is a linear combination of $\mathbf{x}^{(1)}, \ldots, \mathbf{x}^{(\nu)}$, its last $n - \nu$ coordinates are 0. We have

$$|X_\nu^{(\nu)}| \le |\mathbf{X}^{(\nu)}| \ll R_\nu,$$

so we have both the upper bounds in (12.2). The lower bounds follow from a comparison of determinants, together with (12.1). Since $d(\Lambda) = 1$, we have

$$|X_1^{(1)} \cdots X_n^{(n)}| = 1,$$

whence

$$R_1 \cdots R_{\nu-1} |X_\nu^{(\nu)}| R_{\nu+1} \cdots R_n \gg 1,$$

and the right-hand half of (12.1) gives

$$|X_\nu^{(\nu)}| \gg R_\nu.$$

Hence, *a fortiori*, $|\mathbf{X}^{(\nu)}| \gg R_\nu$. Alternatively, the last inequality would follow from the definition of the numbers $R_1, \ldots, R_n$. □

**Note.** The integral basis found in Lemma 12.3 can be further 'normalized', if desired. By adding suitable multiples of $\mathbf{X}^{(1)}, \ldots, \mathbf{X}^{(\nu-1)}$ to $\mathbf{X}^{(\nu)}$ we can ensure that

$$|X_\mu^{(\nu)}| \le \frac{1}{2}|X_\mu^{(\mu)}|$$

for all $\mu, \nu$ with $\mu < \nu$.

**Lemma 12.4.** *Suppose $d(\Lambda) = 1$. Let $N(R)$ denote the number of $\mathbf{x}$ of $\Lambda$ (including the origin) with $|\mathbf{x}| \le R$. Then $N(R) = 1$ if $R < R_1$, and if $R_\nu \le R < R_{\nu+1}$ then*

$$N(R) \asymp \frac{R^\nu}{R_1 R_2 \cdots R_\nu}.$$

**Note.** If $\nu = n$ then $R_{n+1}$ is to be omitted from the condition $R_\nu \le R < R_{\nu+1}$.

*Proof.* The first result stated is obvious; the only point of $\Lambda$ with $|\mathbf{x}| < R_1$ is the origin. To obtain the lower bound for $N(R)$ in the general case, we consider all points of the form

$$\mathbf{x} = u_1\mathbf{X}^{(1)} + \cdots + u_\nu\mathbf{X}^{(\nu)},$$

where $u_1, \ldots, u_\nu$ take all integral values satisfying

$$|u_j| \le \frac{1}{\nu}\frac{R}{|\mathbf{X}^{(j)}|}, \quad (1 \le j \le \nu).$$

All these points have $|\mathbf{x}| \le R$. The number of choices for $u_1, \ldots, u_\nu$ (since zero values are allowed) is

$$\gg \prod_{j=1}^{\nu} \frac{R}{|\mathbf{X}^{(j)}|} \gg \frac{R^\nu}{R_1R_2\cdots R_\nu},$$

by Lemma 12.3.

For the upper bound, we note first that all points $\mathbf{x}$ of $\Lambda$ with $|\mathbf{x}| \le R$ must be linearly dependent on $\mathbf{X}^{(1)}, \ldots \mathbf{X}^{(\nu)}$, since $R < R_{\nu+1}$. Hence they are representable as

$$\mathbf{x} = v_1\mathbf{X}^{(1)} + \cdots + v_\nu\mathbf{X}^{(\nu)}$$

with integers $v_1, \ldots, v_\nu$. For this point, we have

$$\begin{aligned} x_\nu &= v_\nu X_\nu^{(\nu)}, \\ x_{\nu-1} &= v_\nu X_{\nu-1}^{(\nu)} + v_{\nu-1}X_{\nu-1}^{(\nu-1)}, \end{aligned}$$

and so on. Since each coordinate of $\mathbf{x}$ has absolute value $\le R$, the number of possibilities for $v_\nu$ is $\ll R/|X_\nu^{(\nu)}|$, and when $v_\nu$ is chosen, the number of possibilities for $v_{\nu-1}$ is $\ll R/|X_{\nu-1}^{(\nu-1)}|$, and so on. (Note that all these numbers are $\gg 1$, otherwise the argument would be fallacious.) Hence, using Lemma 12.3 again, we conclude that the number of points $\mathbf{x}$ is

$$\ll \frac{R^\nu}{|X_1^{(1)}|\cdots|X_\nu^{(\nu)}|} \ll \frac{R^\nu}{R_1R_2\cdots R_\nu}.$$

$\square$

**Note.** The general meaning of Lemma 12.4 is that, for the purpose of counting the lattice points in a sphere (or other convex body of fixed shape), every lattice behaves like the rectangular lattice generated by

$$(R_1, 0, 0, \ldots, 0), (0, R_2, 0, \ldots, 0), \ldots, (0, 0, 0, \ldots, R_n),$$

up to a constant depending only on $n$. Thus for many purposes we can adequately describe a lattice by means of the $n$ positive numbers $R_1, \ldots, R_n$.

A lattice $\Lambda$ in $x$ space and a lattice $\mathsf{M}$ in $y$ space will be said to be *adjoint*, or *polar* if their bases can be chosen so that

$$\Lambda^T \mathsf{M} = I,$$

where $T$ denotes the transpose of a matrix. If the lattices are given by $\mathbf{x} = \Lambda \mathbf{u}$ and $\mathbf{y} = \mathsf{M}\mathbf{v}$ respectively, where $\mathbf{u}$ and $\mathbf{v}$ are integral vectors the condition means that

$$x_1 y_1 + \cdots + x_n y_n = u_1 v_1 + \cdots + u_n v_n \tag{12.3}$$

identically. Note that $d(\Lambda)d(\mathsf{M}) = 1$. The relation between $\Lambda$ and $\mathsf{M}$ is symmetrical. But the relation is not invariant under arbitrary linear transformations of the $x$ space and the $y$ space; if $\mathbf{x} = A\mathbf{x}'$ and $\mathbf{y} = B\mathbf{y}'$, the relation will only be preserved if $A^T B = I$.

**Lemma 12.5. (Mahler)** *If* $\Lambda, \mathsf{M}$ *are adjoint lattices of determinant* 1, *with successive minima* $R_1, \ldots, R_n$ *and* $S_1, \ldots S_n$, *respectively, then*

$$R_1 \asymp \frac{1}{S_n}, \; R_2 \asymp \frac{1}{S_{n-1}}, \ldots, R_n \asymp \frac{1}{S_1}.$$

*Proof.* Let $\mathbf{x}^{(1)}, \ldots, \mathbf{x}^{(n)}$ be the minimal points for $\Lambda$ (not necessarily a basis) and let $\mathbf{y}^{(1)}, \ldots, \mathbf{y}^{(n)}$ be minimal points for $\mathsf{M}$. The identity (12.3) implies that for any points $\mathbf{x}, \mathbf{y}$ of $\Lambda, \mathsf{M}$ either $\mathbf{x}$ is perpendicular[1] to $\mathbf{y}$ or

$$|x_1 y_1 + \cdots + x_n y_n| \geq 1,$$

whence $|\mathbf{x}||\mathbf{y}| \geq 1$. The $\mathbf{y}$ that are perpendicular to every $\mathbf{x}^{(1)}, \ldots, \mathbf{x}^{(\nu)}$ form an $(n-\nu)$-dimensional linear space. It cannot contain more than $n - \nu$ linearly independent points of $y$ space, and so cannot contain all of $\mathbf{y}^{(1)}, \ldots, \mathbf{y}^{(n-\nu+1)}$. Hence there exist $r \leq \nu$ and $s \leq n - \nu + 1$ such that $|\mathbf{x}^{(r)}||\mathbf{y}^{(s)}| \geq 1$. Now $|\mathbf{x}^{(r)}| \leq R_\nu$ and $|\mathbf{y}^{(s)}| \leq S_{n-\nu+1}$. It follows that

$$R_\nu S_{n-\nu+1} \geq 1, \quad (1 \leq \nu \leq n). \tag{12.4}$$

An inequality in the opposite direction follows by comparing products.

[1] We mean, of course, that the vector from $O$ to $\mathbf{x}$ is perpendicular to that from $O$ to $\mathbf{y}$.

Since

$$R_1 \cdots R_n S_1 \cdots S_n \le \left(\frac{2^n}{J_n}\right)^2$$

by Lemma 12.2, it follows from (12.4) that

$$R_\mu S_{n-\mu+1} \ll \left(\frac{2^n}{J_n}\right)^2, \quad (1 \le \mu \le n).$$

This proves Lemma 12.5. □

**Note.** The condition $d(\Lambda) = 1$ in Lemma 12.5 is unnecessary, but involves no loss of generality. When Lemma 12.5 is extended to the successive minima relative to any convex body (with central symmetry about the origin), it is necessary to use a body in $x$ space and a body in $y$ space which are polar to one another with respect to the unit sphere.

There is a particular type of lattice in $2n$-dimensional space which is essentially self-adjoint. Let $\Lambda$ denote the $2n$-dimensional lattice given by

$$\begin{aligned}
ax_1 &= u_1, \\
&\vdots \\
ax_n &= u_n, \\
a^{-1}x_{n+1} &= \gamma_{11}u_1 + \cdots + \gamma_{1n}u_n + u_{n+1}, \\
&\vdots \\
a^{-1}x_{2n} &= \gamma_{n1}u_1 + \cdots + \gamma_{nn}u_n + u_{2n},
\end{aligned}$$

where $a \ne 0$ and the numbers $\gamma_{ij}$ are real. This has a matrix of the form

$$\Lambda = \begin{pmatrix} a^{-1}I_n & 0 \\ a\gamma & aI_n \end{pmatrix}.$$

The adjoint lattice has the matrix

$$\mathsf{M} = (\Lambda^T)^{-1} = \begin{pmatrix} aI_n & -a\gamma^T \\ 0 & a^{-1}I_n \end{pmatrix}.$$

If $\gamma^T = \gamma$, that is if

$$\gamma_{ij} = \gamma_{ji} \quad \text{for all } i, j, \tag{12.5}$$

then the lattice $\mathsf{M}$ can be transformed into the lattice $\Lambda$ by (i) changing the signs of $v_{n+1}, \ldots, v_{2n}$, (ii) changing the signs of $y_{n+1}, \ldots, y_{2n}$, (iii) interchanging $v_1, \ldots, v_n$ and $v_{n+1}, \ldots, v_{2n}$, (iv) interchanging $y_1, \ldots, y_n$ and $y_{n+1}, \ldots, y_{2n}$.

Hence, subject to (12.5), the successive minima of M are the same as those of $\Lambda$. By the last lemma, it follows that

$$R_1 R_{2n} \asymp 1, \ldots, R_n R_{n+1} \asymp 1.$$

In particular, we have

$$R_n \ll 1 \ll R_{n+1}.$$

We can now prove the main result needed for the later work on cubic forms.

**Lemma 12.6.** *Let $L_1, \ldots, L_n$ be linear forms:*

$$L_i = \gamma_{i1} u_1 + \cdots + \gamma_{in} u_n, \quad (1 \le i \le n),$$

*satisfying the symmetry condition $\gamma_{ij} = \gamma_{ji}$. Let $a > 1$ be real , and let $N(Z)$ denote the number of sets of integers $u_1, \ldots, u_{2n}$ (including 0) satisfying*

$$\begin{cases} |u_1| < aZ, \ldots, |u_n| < aZ, \\ |L_1 - u_{n+1}| < a^{-1}Z, \ldots, |L_n - u_{2n}| < a^{-1}Z. \end{cases} \tag{12.6}$$

*Then, if $0 < Z_1 \le Z_2 \le 1$, we have*

$$\frac{N(Z_2)}{N(Z_1)} \ll \left(\frac{Z_2}{Z_1}\right)^n. \tag{12.7}$$

*Proof.* The inequalities are equivalent to

$$|x_1| < Z, \ldots, |x_n| < Z, \qquad |x_{n+1}| < Z, \ldots, |x_{2n}| < Z$$

for the general point $(x_1, \ldots, x_{2n})$ of the $2n$-dimensional lattice $\Lambda$ defined above. Hence the inequalities imply that $|\mathbf{x}| < \sqrt{2n}Z$. On the other hand, the inequalities are implied by $|\mathbf{x}| < Z$. Hence if $N_0(Z)$ denotes the number of points of $\Lambda$ (including the origin) with $|\mathbf{x}| < Z$, we have

$$N_0(Z) \le N(Z) \le N_0(\sqrt{2n}Z).$$

Thus, if $0 < Z_1 \le Z_2 < 1$,

$$\frac{N(Z_2)}{N(Z_1)} \le \frac{N_0(\sqrt{2n}Z_2)}{N_0(Z_1)}.$$

If we prove the result corresponding to (12.7), namely

$$\frac{N_0(Z_2)}{N_0(Z_1)} \ll \left(\frac{Z_2}{Z_1}\right)^n, \tag{12.8}$$

under the weaker condition $Z_2 \ll 1$ instead of $Z_2 \le 1$, we can apply it with $Z_2$ replaced by $\sqrt{2n}Z_2$ and so deduce (12.7).

Let $R_1, \ldots, R_{2n}$ denote the successive minima of $\Lambda$. We have seen that

$$R_n \ll 1 \ll R_{n+1}.$$

Define $\nu$ and $\mu$ by

$$R_\nu \le Z_1 < R_{\nu+1}, \quad R_\mu \le Z_2 < R_{\mu+1},$$

so that $\nu \le \mu$. By Lemma 12.4,

$$N_0(Z_1) \gg \frac{Z_1^\nu}{R_1 \cdots R_\nu} \quad \text{and} \quad N_0(Z_2) \ll \frac{Z_2^\mu}{R_1 \cdots R_\mu},$$

whence

$$\frac{N_0(Z_2)}{N_0(Z_1)} \ll \frac{Z_2^\mu}{Z_1^\nu R_{\nu+1} \cdots R_\mu}.$$

If $\mu \le n$, the result (12.8) follows, since the right-hand side is

$$\le \frac{Z_2^\mu}{Z_1^\mu} \le \left(\frac{Z_2}{Z_1}\right)^n.$$

If $\mu > n$ and $\nu \le n$, we write the expression as

$$\frac{Z_2^n}{Z_1^\nu R_{\nu+1} \cdots R_n} \frac{Z_2^{\nu-n}}{R_{n+1} \cdots R_\mu},$$

and since $Z_2 \ll 1$ and $R_{n+1} \gg 1$ the result again follows. Finally, the possibility $\nu > n$ can only arise if $Z_1 \gg 1$, in which case

$$\frac{N_0(Z_2)}{N_0(Z_1)} \ll \frac{Z_2^\mu}{Z_1^\nu} \ll 1 \ll \left(\frac{Z_2}{Z_1}\right)^n.$$

This proves Lemma 12.6. □

**Note.** The significance of Lemma 12.6 is that the number of solutions of the inequalities (12.6) does not diminish too rapidly as $Z$ diminishes. The result is of interest only if $aZ$ is large, for if $aZ < 1$ the inequalities imply $u_1 = \cdots = u_{2n} = 0$, and $N(Z) = 1$.

It appears that without the symmetry condition on the coefficients $\gamma_{ij}$ in the linear forms $L_1, \ldots, L_n$, one could assert only a weaker result in which the exponent $n$ would be replaced by $2n - 1$.

# 13

# Cubic forms

We now set out to prove that a homogeneous cubic equation

$$C(x_1, \ldots, x_n) = 0, \tag{13.1}$$

with integral coefficients, is always soluble in integers $x_1, \ldots, x_n$ (not all 0) if $n \geq 17$. The first such result, with the condition $n \geq 32$, was proved in 1957 [21], and the improved result was found early in 1962 [20]. In 1963 I proved that the condition $n \geq 16$ suffices [22], but this requires a more detailed argument of a somewhat special nature, beyond what is needed for 17.

It was pointed out by Mordell [61] in 1937 that there exist cubic forms in nine variables which do not represent zero, and consequently the condition $n \geq 10$ is essential if (13.1) is to be always soluble. The example of Mordell is based on the properties of a norm form of a cubic field. If $p$ is a prime which does not factorize in the field, then the norm form $N(x, y, z)$ is never divisible by $p$ except when $x$, $y$, $z$ are all divisible by $p$. It follows easily that the equation

$$N(x_1, x_2, x_3) + pN(x_4, x_5, x_6) + p^2 N(x_7, x_8, x_9) = 0$$

has no solution in integers $x_1, \ldots, x_9$ except the trivial solution. Indeed, we could assert further that the corresponding congruence to the modulus $p^3$ has no solution except with all the variables 0 (mod $p$). A simple example would be provided by taking $p = 7$ and

$$N(x, y, z) = x^3 + 2y^3 + 4z^3 - 6xyz,$$

this being the norm form of the field generated by $\sqrt[3]{2}$. A similar construction to that above gives examples of homogeneous equations of degree $k$ in $k^2$ variables which are insoluble.

The proof of the theorem on cubic equations falls into several chapters, each of which is largely self-contained. We begin by considering the exponential sum associated with a cubic form. Write

$$C(\mathbf{x}) = C(x_1, \dots, x_n) = \sum_i \sum_j \sum_k c_{ijk} x_i x_j x_k,$$

where the summations go from 1 to $n$. The coefficients $c_{ijk}$ are integers, and we can suppose that $c_{ijk}$ is a symmetrical function of $i$, $j$, $k$. Let $P$ be a large positive integer. Let $\mathfrak{B}$ be a fixed box in $n$ dimensional space, namely the cartesian product of $n$ intervals

$$x_j' < x_j \le x_j'', \quad (1 \le j \le n).$$

We shall suppose, merely for convenience, that $x_j'' - x_j' < 1$. Let

$$S(\alpha) = \sum_{P\mathfrak{B}} e\left(\alpha C(x_1, \dots, x_n)\right),$$

where the summation is over all integer points in the box $P\mathfrak{B}$, given by

$$Px_j' < x_j \le Px_j'', \quad (1 \le j \le n).$$

Let $\mathcal{N}(P)$ denote the number of integer points $\mathbf{x}$ in $P\mathfrak{B}$ which satisfy (13.1). Then

$$\mathcal{N}(P) = \int_0^1 S(\alpha)\, d\alpha. \tag{13.2}$$

Our aim (in principle) is to prove that, with a suitable choice of the box $\mathfrak{B}$, there is an asymptotic formula for $\mathcal{N}(P)$ as $P \to \infty$, in which the main term is of order $P^{n-3}$. Actually this is not always true. What we *shall* arrive at is the apparently paradoxical result that the asymptotic formula holds on the hypothesis that (13.1) is insoluble (so that $\mathcal{N}(P) = 0$)! This will suffice for our purpose, because it will prove that (13.1) is soluble.

The trivial estimate for $|S(\alpha)|$ is $P^n$. We begin by investigating what happens if, for some particular $\alpha$,

$$|S(\alpha)| \ge P^{n-K}, \tag{13.3}$$

where $K$ is some positive number. Ultimately our aim, as in earlier chapters, is to be able to remove from the set of $\alpha$ any subset which contributes an amount of lower order than $P^{n-3}$ to the integral.

The first step is to prove a generalization of Weyl's inequality. We define, for any two points $\mathbf{x}$, $\mathbf{y}$, a set of $n$ bilinear forms:

$$B_j(\mathbf{x} \,|\, \mathbf{y}) = \sum_i \sum_k c_{ijk} x_i y_k, \quad (1 \le j \le n).$$

**Lemma 13.1.** *The hypothesis (13.3) implies that*

$$\sum_{|\mathbf{x}|<P} \sum_{|\mathbf{y}|<P} \prod_{j=1}^{n} \min\left(P, \|6\alpha B_j(\mathbf{x}\,|\,\mathbf{y})\|^{-1}\right) \gg P^{3n-4K}.$$

*Proof.* We have

$$\begin{aligned} |S(\alpha)|^2 &= \sum_{\mathbf{z}\in P\mathfrak{B}} \sum_{\mathbf{z}'\in P\mathfrak{B}} e\left(\alpha C(\mathbf{z}') - \alpha C(\mathbf{z})\right) \\ &= \sum_{\mathbf{z}\in P\mathfrak{B}} \sum_{\mathbf{y}+\mathbf{z}\in P\mathfrak{B}} e\left(\alpha C(\mathbf{y}+\mathbf{z}) - \alpha C(\mathbf{z})\right). \end{aligned}$$

For any $\mathbf{z}$, the box $P\mathfrak{B} - \mathbf{z}$ is contained in $|\mathbf{y}| < P$. Hence

$$|S(\alpha)|^2 \le \sum_{|\mathbf{y}|<P} \left| \sum_{\mathbf{z}\in\mathcal{R}(\mathbf{y})} e\left(\alpha C(\mathbf{y}+\mathbf{z}) - \alpha C(\mathbf{z})\right) \right|,$$

where $\mathcal{R}(\mathbf{y})$ denotes the common part of the boxes $P\mathfrak{B}$ and $P\mathfrak{B} - \mathbf{y}$. By Cauchy's inequality,

$$|S(\alpha)|^4 \ll P^n \sum_{|\mathbf{y}|<P} \left| \sum_{\mathbf{z}\in\mathcal{R}(y)} e\left(\alpha C(\mathbf{y}+\mathbf{z}) - \alpha C(\mathbf{z})\right) \right|^2.$$

We now repeat the argument on the inner sum. Its square does not exceed

$$\sum_{|\mathbf{x}|<P} \left| \sum_{\mathbf{z}\in\mathcal{S}(\mathbf{x},\mathbf{y})} e\left(\alpha C(\mathbf{z}+\mathbf{x}+\mathbf{y}) - \alpha C(\mathbf{z}+\mathbf{x}) - \alpha C(\mathbf{z}+\mathbf{y}) + \alpha C(\mathbf{z})\right) \right|,$$

where $\mathcal{S}(\mathbf{x},\mathbf{y})$ is a box for $\mathbf{z}$, depending on $\mathbf{x}$ and $\mathbf{y}$, with edges less than $P$ in length. We have

$$\begin{aligned} &\alpha(C(\mathbf{z}+\mathbf{x}+\mathbf{y}) - C(\mathbf{z}+\mathbf{x}) - C(\mathbf{z}+\mathbf{y}) + C(\mathbf{z})) \\ &\quad = 6\alpha \sum_{i,j,k} c_{ijk} x_i y_k z_j + \phi = 6\alpha \sum_j z_j B_j(\mathbf{x}\,|\,\mathbf{y}) + \phi, \end{aligned}$$

where $\phi$ does not involve $z$. By a now familiar estimate,

$$\left| \sum_{\mathbf{z}\in\mathcal{S}(\mathbf{x},\mathbf{y})} e\left(6\alpha \sum_j B_j(\mathbf{x}\,|\,\mathbf{y}) z_j\right) \right| \ll \prod_{j=1}^{n} \min\left(P, \|6\alpha B_j(\mathbf{x}\,|\,\mathbf{y})\|^{-1}\right).$$

Substitution in the previous inequalities, together with (13.3), yields the result. □

**Note.** It is a useful precaution, when estimating an exponential sum, to see what the trivial estimate yields, in order to judge how much (if anything) has been lost for ever. In the present case, taking $P$ for the minimum throughout the product, the trivial estimate for $|S(\alpha)|^4$ would be $P^{4n}$, which is satisfactory.

**Lemma 13.2.** *The hypothesis (13.3) implies that the number of pairs* $\mathbf{x}$, $\mathbf{y}$ *of integer points that satisfy*[1]

$$|\mathbf{x}| < P, \quad |\mathbf{y}| < P, \quad \|6\alpha B_j(\mathbf{x}\,|\,\mathbf{y})\| < P^{-1}, \quad (1 \le j \le n), \qquad (13.4)$$

*is*

$$\gg P^{2n-4K}(\log P)^{-n}.$$

*Proof.* Let $Y(\mathbf{x})$ denote the number of points $\mathbf{y}$ satisfying (13.4) for given $\mathbf{x}$. Then, for any integers $r_1, \ldots, r_n$ with $0 \le r_j < P$, there cannot be more than $Y(\mathbf{x})$ integer points $\mathbf{y}$, with each coordinate in some prescribed interval of length $P$, which satisfy

$$\frac{r_j}{P} \le \{6\alpha B_j(\mathbf{x}\,|\,\mathbf{y})\} < \frac{r_j+1}{P}, \quad (1 \le j \le n),$$

where $\{\theta\}$ denotes the fractional part of any real number $\theta$. For if $\mathbf{y}'$ were one such point, and $\mathbf{y}$ were any such point, we should have $|\mathbf{y}-\mathbf{y}'| < P$ and

$$\|6\alpha B_j(\mathbf{x}\,|\,\mathbf{y}-\mathbf{y}')\| < P^{-1}, \quad (1 \le j \le n).$$

Thus there cannot be more than $Y(\mathbf{x})$ possibilities for $\mathbf{y}$. (Note that $\mathbf{y} = \mathbf{0}$ is permitted in (13.4).)

Dividing the cube $|\mathbf{y}| < P$ into $2^n$ cubes of side $P$, we obtain

$$\begin{aligned} &\sum_{|\mathbf{y}|<P} \prod_{j=1}^{n} \min(P, \|6\alpha B_j(\mathbf{x}\,|\,\mathbf{y})\|^{-1}) \\ &\ll Y(\mathbf{x}) \sum_{r_1=0}^{P-1} \cdots \sum_{r_n=0}^{P-1} \min\left(P, \frac{P}{r_j}, \frac{P}{r_j-1}\right) \\ &\ll Y(\mathbf{x})(P\log P)^n. \end{aligned}$$

[1] Here we put $|\mathbf{x}| = \max(|x_1|, \ldots, |x_n|)$ for any point $\mathbf{x}$. This is a different notation from that of Chapter 12 where $|\mathbf{x}|$ denotes the distance of the point $\mathbf{x}$ from the origin. For our purposes, however, the difference is unimportant.

Substitution in the result of Lemma 13.1 gives

$$(P \log P)^n \sum_{|\mathbf{x}|<P} Y(\mathbf{x}) \gg P^{3n-4K}.$$

Since $\sum Y(\mathbf{x})$ is the number of pairs $\mathbf{x}$, $\mathbf{y}$ satisfying (13.4), the result follows. □

**Note.** Since the trivial estimate for the number of pairs $\mathbf{x}$, $\mathbf{y}$ satisfying (13.4) is $P^{2n}$, we have abandoned a factor $(\log P)^n$. But this is not important.

**Lemma 13.3.** *Let $\theta$ be independent of $P$ and satisfy $0 < \theta < 1$. The hypothesis (13.3) implies that the number of pairs $\mathbf{x}$, $\mathbf{y}$ of integer points satisfying*

$$|\mathbf{x}| < P^{\theta}, \quad |\mathbf{y}| < P^{\theta}, \quad \|6\alpha B_j(\mathbf{x} \,|\, \mathbf{y})\| < P^{-3+2\theta} \tag{13.5}$$

*is*

$$\gg P^{2n\theta-4K}(\log P)^{-n}.$$

*Proof.* We shall get the result by two applications of Lemma 12.6. First we look upon $\mathbf{x}$ as fixed in (13.4), and consider the number of integer points $\mathbf{y}$. The inequalities for $\mathbf{y}$ are

$$\begin{cases} |y_1| < P, \ldots, |y_n| < P, \\ |L_1(\mathbf{y}) - u_{n+1}| < P^{-1}, \ldots, |L_n(\mathbf{y}) - u_{2n}| < P^{-1}, \end{cases} \tag{13.6}$$

where $L_j(\mathbf{y}) = 6\alpha B_j(\mathbf{x} \,|\, \mathbf{y})$, and where $u_{n+j}$ is the integer nearest to $L_j(\mathbf{y})$. The forms $L_j(\mathbf{y})$ satisfy the symmetry condition of the preceding chapter, since the coefficient of $y_k$ in $L_j$ is $6\alpha \sum_i c_{ijk} x_i$, and this is unaltered by interchanging $j$ and $k$. We apply Lemma 12.6 with $y_1, \ldots, y_n$ for $u_1, \ldots, u_n$ and with

$$a = P, \quad Z_2 = 1, \quad Z_1 = P^{-1+\theta}.$$

When $Z_2 = 1$, the inequalities of Lemma 12.6 are the inequalities (13.6) above. Suppose these have $N(\mathbf{x})$ solutions in $\mathbf{y}$. The inequalities of Lemma 12.6 with $Z = Z_1$ become

$$|y_1| < P^{\theta}, \ldots, |y_n| < P^{\theta},$$
$$|L_1(\mathbf{y}) - u_{n+1}| < P^{-2+\theta}, \ldots, |L_n(\mathbf{y}) - u_{2n}| < P^{-2+\theta}.$$

Hence the number of solutions of these in $\mathbf{y}$ is

$$\gg N(\mathbf{x}) P^{-n(1-\theta)}.$$

Hence the number of pairs $\mathbf{x}$, $\mathbf{y}$ which satisfy

$$|\mathbf{x}| < P, \quad |\mathbf{y}| < P^{\theta}, \quad \|6\alpha B_j(\mathbf{x}\,|\,\mathbf{y})\| < P^{-2+\theta} \tag{13.7}$$

is

$$\gg P^{-n(1-\theta)} \sum_{|\mathbf{x}|<P} N(\mathbf{x}) \gg P^{n+n\theta-4K}(\log P)^{-n},$$

by Lemma 13.2.

Now we go through a similar argument, but with $\mathbf{y}$ fixed and $\mathbf{x}$ variable. For each $\mathbf{y}$, the conditions (13.7) on $\mathbf{x}$ are

$$|x_1| < P, \dots, |x_n| < P,$$
$$|M_1(\mathbf{x}) - u_{n+1}| < P^{-2+\theta}, \dots, |M_n(\mathbf{x}) - u_{2n}| < P^{-2+\theta},$$

where $M_j(\mathbf{x}) = B_j(\mathbf{y}\,|\,\mathbf{x})$ and $u_{n+j}$ is now the integer nearest to $M_j(\mathbf{x})$. These are the inequalities of Lemma 12.6 with

$$a = P^{\frac{3}{2}-\frac{1}{2}\theta}, \quad Z = Z_2 = P^{-\frac{1}{2}+\frac{1}{2}\theta}.$$

We take $Z_1 = P^{-\frac{3}{2}+\frac{3}{2}\theta}$. Then the lemma tells us that the number of solutions of

$$|x_1| < P^{\theta}, \dots, |x_n| < P^{\theta},$$
$$|M_1(\mathbf{x}) - u_{n+1}| < P^{-3+2\theta}, \dots, |M_n(\mathbf{x}) - u_{2n}| < P^{-3+2\theta},$$

is $\gg P^{-n(1-\theta)} N_1(\mathbf{y})$, where $N_1(\mathbf{y})$ denotes the number of solutions of (13.7) in $\mathbf{x}$ for given $\mathbf{y}$. Hence the number of pairs of integer points satisfying (13.5) is

$$\gg P^{-n+n\theta} \sum_{|\mathbf{y}|<P^{\theta}} N_1(\mathbf{y}) \gg P^{-n+n\theta} P^{n+n\theta-4K}(\log P)^{-n},$$

whence the result. □

**Lemma 13.4.** *Let $\theta$ be independent of $P$ and satisfy $0 < \theta < 1$. Let $\varepsilon$ be any small fixed positive number. Then either*

(A) *there are more than $P^{n\theta+\varepsilon}$ pairs $\mathbf{x}$, $\mathbf{y}$ of integer points satisfying*

$$|\mathbf{x}| < P^{\theta}, \quad |\mathbf{y}| < P^{\theta}, \quad B_j(\mathbf{x}\,|\,\mathbf{y}) = 0, \quad (1 \le j \le n), \tag{13.8}$$

*or*

(B) *for every $\alpha$ the hypothesis*

$$|S(\alpha)| \ge P^{n-\frac{1}{4}n\theta+\varepsilon} \tag{13.9}$$

*implies that $\alpha$ has a rational approximation $a/q$ such that*

$$(a, q) = 1, \quad 1 \le q \ll P^{2\theta}, \quad |q\alpha - a| < P^{-3+2\theta}. \tag{13.10}$$

*Proof.* We take $K = \frac{1}{4}n\theta - \varepsilon$ in (13.3), so that this becomes the same as (13.9). By Lemma 13.3, there are

$$\gg P^{2n\theta - 4K}(\log P)^{-n} \gg P^{n\theta+\varepsilon}$$

pairs $\mathbf{x}$, $\mathbf{y}$ of integer points satisfying (13.5). If, for all points, $B_j(\mathbf{x} \mid \mathbf{y}) = 0$ for all $j$, we have alternative (A) active. If not, then for some pair $\mathbf{x}$, $\mathbf{y}$ in (13.5) and some $j$, we have $B_j(\mathbf{x} \mid \mathbf{y}) \ne 0$ and

$$\|6\alpha B_j(\mathbf{x} \mid \mathbf{y})\| < P^{-3+2\theta}.$$

Take $q = 6|B_j(\mathbf{x} \mid \mathbf{y})|$ and take $a$ to be the nearest integer to $\alpha q$. Then

$$|q\alpha - a| < P^{-3+2\theta}.$$

Also

$$\begin{aligned} q \ll |B_j(\mathbf{x} \mid \mathbf{y})| &\ll \sum_{i,k} |c_{ijk}||x_i||y_k| \\ &\ll |\mathbf{x}||\mathbf{y}| \ll P^{2\theta}. \end{aligned}$$

We do not necessarily have $(a, q) = 1$, but this can be ensured by removing any common factor from $q$ and $a$. Thus we obtain alternative (B). □

**Note.** It will be seen that alternative A does not involve $\alpha$. Nor does it involve $\theta$ essentially, for if we put $R = P^{\theta}$, the assertion is that the number of pairs $\mathbf{x}$, $\mathbf{y}$ satisfying

$$|\mathbf{x}| < R, \quad |\mathbf{y}| < R, \quad B_j(\mathbf{x} \mid \mathbf{y}) = 0, \quad (1 \le j \le n), \tag{13.11}$$

is greater than $R^{n+\varepsilon'}$ for some fixed $\varepsilon' > 0$. Thus alternative A relates to an intrinsic property of the cubic form. Note that we could exclude $\mathbf{x} = \mathbf{0}$ and $\mathbf{y} = \mathbf{0}$ from (13.11) if we wished, since the number of such pairs is $\ll R^n$.

Alternative B gives us a situation which is similar in principle to that with which we have become familiar in earlier chapters. It will enable us, in Chapter 15, to estimate satisfactorily the contribution made to $\int_0^1 S(\alpha)\, d\alpha$ by a large set of $\alpha$ and leave us with a relatively small number of short intervals in which we can approximate to $S(\alpha)$.

# 14

# Cubic forms: bilinear equations

We now investigate alternative A of Lemma 13.4, namely that for some arbitrarily large $R$ there exist more than $R^{n+\varepsilon}$ pairs $\mathbf{x}, \mathbf{y}$ of integer points satisfying $0 < |\mathbf{x}| < R$, $0 < |\mathbf{y}| < R$ and

$$B_j(\mathbf{x} \,|\, \mathbf{y}) = 0, \quad (1 \le j \le n). \tag{14.1}$$

We have excluded $\mathbf{x} = \mathbf{0}$ and $\mathbf{y} = \mathbf{0}$, in accordance with the remark just made. In this chapter, which is self-contained, we shall prove that this implies the existence of an integer point $\mathbf{z} \neq \mathbf{0}$ for which $C(\mathbf{z}) = 0$. Actually a slightly weaker hypothesis would suffice, namely that there are more than $AR^n$ pairs $\mathbf{x}, \mathbf{y}$, where $A$ is greater than some function of $n$. But the use of $\varepsilon$ simplifies the exposition.

For any particular $\mathbf{x}$, the equations (14.1) are $n$ linear equations in $\mathbf{y} = (y_1, \ldots, y_n)$, and their determinant is

$$H(\mathbf{x}) = \det\left(\sum_{i=1}^{n} c_{ijk} x_i\right), \quad (1 \le j, k \le n). \tag{14.2}$$

This is the Hessian of the cubic form $C(\mathbf{x})$. It is a form of degree $n$ in $\mathbf{x}$; or one should rather say of apparent degree $n$, since it may identically vanish. We must first prove that this cannot happen if $C(\mathbf{x})$ does not represent zero.

**Lemma 14.1.** *If $C(\mathbf{x}) \neq 0$ for all integral $\mathbf{x} \neq \mathbf{0}$, then $H(\mathbf{x})$ does not vanish identically.*

*Proof.* Suppose $H(\mathbf{x}) = 0$ identically. Let $n - r$ $(r \ge 1)$ be the identical rank of $H(\mathbf{x})$; that is, suppose all subdeterminants of the matrix in (14.2) of order $n - r + 1$ vanish identically in $\mathbf{x}$ but some subdeterminant of order $n - r$ does not. Suppose, for convenience of exposition, that the last determinant, say $\Delta$, is in the top left-hand corner. Then the first $n-r$ of the equations (14.1) imply all the others, since the later rows of the matrix are linearly dependent on the first $n-r$ rows. Let $\Delta_j$ denote the determinant obtained from $\Delta$ by replacing the $j$th column by the $(n-r+1)$th column. Then a particular solution of the equations $B_j(\mathbf{x}\,|\,\mathbf{y}) = 0$ is given by

$$y_1 = \Delta_1, \ldots, y_{n-r} = \Delta_{n-r}, y_{n-r+1} = -\Delta, y_{n-r+2} = \cdots = 0.$$

(This follows from Cramer's rule: we solve the first $n-r$ equations, with $y_{n-r+1} = 1$, $y_{n-r+2} = 0, \ldots$, and multiply throughout by $-\Delta$.)

In this solution, $y_1, \ldots, y_n$ are forms in $x_1, \ldots, x_n$ with integral coefficients, which do not all vanish identically, since $\Delta$ does not. We have

$$B_j(\mathbf{x}\,|\,\mathbf{y}) = \sum_{i,k} c_{ijk} x_i y_k = 0$$

identically in $x_1, \ldots, x_n$. We now regard $x_1, \ldots, x_n$ as continuous variables, and differentiate this identity with respect to any $x_\nu$, getting

$$\sum_k c_{\nu jk} y_k + \sum_{i,k} c_{ijk} x_i \frac{\partial y_k}{\partial x_\nu} = 0$$

for all $j$ and $\nu$. Multiply by $y_j$ and sum over $j$, and note that

$$\sum_i \sum_j c_{ijk} x_i y_j = 0$$

for all $k$. We get

$$\sum_{j,k} c_{\nu jk} y_j y_k = 0$$

for all $\nu$. This implies, in particular, that $C(\mathbf{y}) = 0$. If we now take $\mathbf{x}$ to be any integer point for which $\mathbf{y} \neq \mathbf{0}$ (as is possible because $\mathbf{y}$ is not identically $\mathbf{0}$) we get a contradiction to the hypothesis. This proves the lemma. □

The last lemma shows that the points $\mathbf{x}$, for which there is a non-zero solution of the linear equations $B_j(\mathbf{x}\,|\,\mathbf{y}) = 0$ in $\mathbf{y}$, satisfy the non-identical equation $H(\mathbf{x}) = 0$. Thus the number of such integral points with $0 < |\mathbf{x}| < R$ is $\ll R^{n-1}$. We next extend this result by proving that, for $r = 1, \ldots, n-1$, the number of integral points $\mathbf{x}$ with $0 < |\mathbf{x}| < R$ for which the equations have $r$ linearly independent solutions in $\mathbf{y}$ is $\ll R^{n-r}$. The result already proved is the case $r = 1$.

It is convenient to deal first with a question of elementary algebraic geometry.

**Lemma 14.2.** *Let*

$$f_1(\mathbf{x}), \ldots, f_N(\mathbf{x})$$

*be forms with integral coefficients in* $\mathbf{x}_1, \ldots, \mathbf{x}_n$ *and suppose that* $N$ *and the (common) degree of* $f_1, \ldots, f_N$ *are bounded in terms of* $n$. *Suppose that for some arbitrarily large* $R$, *there is a set* $\mathcal{X}$ *of integer points* $\mathbf{x}$ *satisfying*

$$0 < |\mathbf{x}| < R, \qquad f_1(\mathbf{x}) = 0, \ldots, f_N(\mathbf{x}) = 0,$$

*and suppose that there are more than* $R^{n-r+\varepsilon}$ *points in* $\mathcal{X}$. *Then, for some one of the points* $\mathbf{x}$ *of* $\mathcal{X}$, *there exist numbers* $T_{i\rho}$, $D_{\rho\nu}$, *such that*

$$\frac{\partial f_i}{\partial x_\nu} = \sum_{\rho=1}^{r-1} T_{i\rho} D_{\rho\nu},$$

*for* $i = 1, \ldots, N$ *and* $\nu = 1, \ldots, n$, *at that point.*

**Note.** The equations for $\partial f_i / \partial x_\nu$ are equivalent to the assertion that the rank of the matrix $\partial f_i / \partial x_\nu$ for $i = 1, \ldots, N$ and $\nu = 1, \ldots, n$, is $\leq r - 1$. This follows from elementary matrix theory.

*Proof.* The equations $f_1(\mathbf{x}) = 0, \ldots, f_N(\mathbf{x}) = 0$ define an algebraic variety in $n$-dimensional space (or $(n-1)$-dimensional projective space). Such a variety is expressible as a union of absolutely irreducible varieties, and the number of these, under the present hypotheses, is bounded in terms of $n$. Hence there is one of these absolutely irreducible varieties, say $\mathcal{V}$, which contains more than $R^{n-r+\frac{1}{2}\varepsilon}$ points of $\mathcal{X}$.

Associated with the absolutely irreducible variety $\mathcal{V}$, considered as a set of points in complex space, is its *dimension* $s$. We need only the following property of $s$: the irreducible variety $\mathcal{V}$ can be decomposed into a bounded number of parts, such that, on each part, $s$ of $x_1, \ldots, x_n$ are independent variables and the other $n - s$ are single valued differentiable functions of them.

It follows that $\mathcal{V}$ contains $\ll R^s$ integer points $\mathbf{x}$ satisfying $|\mathbf{x}| < R$, since there are $\ll R^s$ possibilities for each of the $s$ coordinates. Comparison with the earlier statement about the number of points of $\mathcal{X}$ on $\mathcal{V}$ shows that

$$s \geq n - r + 1.$$

Now consider the neighbourhood of any one point of $\mathcal{X}$ on $\mathcal{V}$. We can suppose that here $x_{s+1}, \ldots, x_n$ are single valued and differentiable functions of $x_1, \ldots, x_s$. Let $f(x_1, \ldots, x_n)$ be any differentiable function which vanishes everywhere on $\mathcal{V}$. Then, differentiating the identity $f = 0$ with respect to $x_\nu$, for $\nu = 1, \ldots, s$, we get

$$f^{(\nu)} + f^{(s+1)} \frac{\partial x_{s+1}}{\partial x_\nu} + \cdots + f^{(n)} \frac{\partial x_n}{\partial x_\nu} = 0,$$

where $f^{(j)} = \partial f / \partial x_j$ when $x_1, \ldots, x_n$ are independent variables. Thus for $\nu = 1, \ldots, s$ we have

$$\begin{aligned} f^{(\nu)} &= \sum_{\rho=1}^{n-s} f^{(s+\rho)} \left( -\frac{\partial x_{s+\rho}}{\partial x_\nu} \right) \\ &= \sum_{\rho=1}^{n-s} f^{(s+\rho)} D_{\rho,\nu}, \end{aligned}$$

say. If we define $D_{\rho,\nu}$ for $\nu > s$ by

$$D_{\rho,\nu} = \begin{cases} 1 & \text{if } s + \rho = \nu, \\ 0 & \text{otherwise,} \end{cases}$$

the same relations holds for $\nu = s+1, \ldots, n$. Hence

$$f^{(\nu)}(\mathbf{x}) = \sum_{\rho=1}^{n-s} T_\rho(f) D_{\rho,\nu}$$

for $\nu = 1, \ldots, n$, where $T_\rho(f) = f^{(s+\rho)}$. Note that the numbers $D_{\rho,\nu}$ are independent of $f$, and the numbers $T_\rho(f)$ are independent of $\nu$. This proves the relations in the enunciation, since $n - s \leq r - 1$, and they hold at any point on $\mathcal{V}$, and in particular at any of the points of $\mathcal{X}$ on $\mathcal{V}$. □

**Lemma 14.3.** *Suppose $C(\mathbf{x}) \neq 0$ for all integral $\mathbf{x} \neq \mathbf{0}$. Then the number of integer points $\mathbf{x}$ with $0 < |\mathbf{x}| < R$ for which the bilinear equations $B_j(\mathbf{x} \,|\, \mathbf{y}) = 0$ have exactly $r$ linearly independent solutions in $\mathbf{y}$ is less than $R^{n-r+\varepsilon}$.*

*Proof.* The points in question are those for which the rank of the matrix $(\sum_i c_{ijk} x_i)$ is exactly $r$. It will suffice to consider the set $\mathcal{X}$ of integer points for which some particular subdeterminant of order $n - r$ is $\neq 0$ and all subdeterminants of order $n - r + 1$ are 0, and to prove that the number of points in $\mathcal{X}$ is less than $R^{n-r+\varepsilon}$. Suppose the number of points in $\mathcal{X}$ is $\geq R^{n-r+\varepsilon}$. Then we have the situation of Lemma 14.2, where $f_1(\mathbf{x}), \ldots, f_N(\mathbf{x})$ are all the subdeterminants of order $n-r+1$.

For any points $\mathbf{x}$ of $\mathcal{X}$ we can construct $r$ linearly independent solutions $\mathbf{y}^{(1)}, \ldots, \mathbf{y}^{(r)}$ of the bilinear equations, as in the proof of Lemma 14.1, by taking the coordinates of these points to be certain subdeterminants of order $n-r$. (In that proof, we needed only one solution, which we got by taking $y_{n-r+1} = -\Delta$, $y_{n-r+2} = \cdots = 0$; but the extension to $r$ solutions is immediate.)

Now consider the values of $B_j(\mathbf{x} \,|\, \mathbf{y}^{(p)})$ when $\mathbf{x}$ is an arbitrary point (real or complex) and $\mathbf{y}^{(p)}$ is as above. This value will be a certain subdeterminant of order $n-r+1$ of the matrix mentioned at the beginning. (Sometimes it will be a subdeterminant of order $n-r+1$ with two identical rows, but that is immaterial.) Hence for *any* point $\mathbf{x}$ we have the identities

$$\sum_{i,k} c_{ijk} x_i y_k^{(p)} = \Delta_{j,p}(\mathbf{x}),$$

where $\Delta_{j,p}(\mathbf{x})$ is some subdeterminant of order $n-r+1$. Of course, all the $\Delta_{j,p}$ vanish if $\mathbf{x}$ is in $\mathcal{X}$.

In the above identities, $x_1, \ldots, x_n$ are independent variables. Differentiation with respect to $x_\nu$ gives

$$\sum_k c_{\nu jk} y_k^{(p)} + \sum_{i,k} c_{ijk} x_i \frac{\partial y_k^{(p)}}{\partial x_\nu} = \Delta_{j,p}^{(\nu)}(\mathbf{x}),$$

where the superscript $(\nu)$ on the right denotes a partial derivative. Multiply by $y_j^{(q)}$ $(1 \leq q \leq r)$ and sum over $j$. We get

$$\sum_{j,k} c_{\nu jk} y_k^{(p)} y_j^{(q)} + \sum_k \Delta_{k,q} \frac{\partial y_k^{(p)}}{\partial x_\nu} = \sum_j y_j^{(q)} \Delta_{j,p}^{(\nu)}(\mathbf{x}).$$

Now consider any point $\mathbf{Y}$ in the linear space of $r$ dimensions generated by $\mathbf{y}^{(1)}, \ldots, \mathbf{y}^{(r)}$, say

$$\mathbf{Y} = \sum_{p=1}^{r} K_p \mathbf{y}^{(p)}.$$

For this point, we have

$$\sum_{j,k} c_{\nu jk} Y_j Y_k + \sum_{p=1}^{r} \sum_{q=1}^{r} K_p K_q \sum_k \Delta_{k,q} \frac{\partial y_k^{(p)}}{\partial x_\nu}$$

$$= \sum_{p=1}^{r}\sum_{q=1}^{r} K_p K_q \sum_j y_j^{(q)} \Delta_{j,p}^{(\nu)}$$

$$= \sum_{p=1}^{r}\sum_j A_{j,p}\Delta_{j,p}^{(\nu)},$$

say. This holds for $\nu = 1, \ldots, n$.

Take the point $\mathbf{x}$ to be one of the points found in Lemma 14.2. For this point, which is fixed from now onwards, we have $\Delta_{k,q} = 0$ for all $k$ and $q$, and

$$\Delta_{j,p}^{(\nu)} = \sum_{\rho=1}^{r-1} T_{j,p,\rho} D_{\rho,\nu}.$$

Hence

$$\sum_{j,k} c_{\nu jk} Y_j Y_k = \sum_{p=1}^{r}\sum_j A_{j,p} \sum_{\rho=1}^{r-1} T_{j,p,\rho} D_{\rho,\nu}.$$

Finally, multiply by $Y_\nu = \sum_{\sigma=1}^{r} K_\sigma y_\nu^{(\sigma)}$ and sum over $\nu$. We get

$$C(\mathbf{Y}) = \sum_{p=1}^{r}\sum_j A_{j,p} \sum_{\sigma=1}^{r}\sum_{\rho=1}^{r-1} K_\sigma y_\nu^{(\sigma)} T_{j,p,\rho} D_{\rho,\nu}.$$

Now choose the numbers $K_1, \ldots, K_r$ to satisfy

$$\sum_{\sigma=1}^{r} K_\sigma \sum_\nu y_\nu^{(\sigma)} D_{\rho,\nu} = 0, \quad (1 \le \rho \le r-1).$$

These are $r-1$ linear equations in $r$ unknowns, and so have a solution with $K_1, \ldots, K_r$ not all 0. Also, since the numbers $D_{\rho,\nu}$ can be supposed rational, we can take $K_1, \ldots, K_r$ to be integers. Hence $\mathbf{Y}$ is an integer point not $\mathbf{0}$, because $\mathbf{y}^{(1)}, \ldots, \mathbf{y}^{(r)}$ are linearly independent since the point $\mathbf{x}$ is in $\mathcal{X}$. We get $C(\mathbf{Y}) = 0$, contrary to hypothesis, and this proves the result. □

Note that in the above proof, the choice of $K_1, \ldots, K_r$ did not involve the numbers $A_{j,p}$. Had it done so, the reasoning would have been fallacious, since these themselves depend on $K_1, \ldots, K_r$.

**Lemma 14.4.** *Alternative A of Lemma 13.4 implies that $C(\mathbf{x})$ represents 0.*

*Proof.* If $C(\mathbf{x})$ does not represent zero then Lemma 14.1 and Lemma 14.3

imply that there are $\ll R^{n-r+\varepsilon}$ points $\mathbf{x}$ such that there are exactly $r$ linearly independent solutions of the bilinear equations in $\mathbf{y}$. Hence there are $\ll R^{n+\varepsilon}$ pairs $\mathbf{x}, \mathbf{y}$ with $0 < |\mathbf{x}| < R$, $0 < |\mathbf{y}| < R$ which satisfy the bilinear equations. But this contradicts alternative A, since we can take the present $\varepsilon$ to be (say) half of the $\varepsilon$ in alternative A. (Actually the present proof shows that the number of pairs $\mathbf{x}, \mathbf{y}$ is $\ll R^n$, as remarked earlier.) $\square$

# 15

# Cubic forms: minor arcs and major arcs

From now onwards we can suppose that alternative B of Lemma 13.4 holds. Thus, for any $\alpha$, either

$$|S(\alpha)| < P^{n-\frac{1}{4}n\theta+\varepsilon} \tag{15.1}$$

or $\alpha$ lies in the set, which we shall call $\xi(\theta)$, of real numbers which have a rational approximation $a/q$ satisfying

$$(a,q)=1, \quad 1 \leq q \ll P^{2\theta}, \quad |q\alpha - a| \ll P^{-3+2\theta}. \tag{15.2}$$

We shall find that, provided $n \geq 17$, this result enables us to estimate satisfactorily the contribution made to the integral for $\mathcal{N}(P)$by all $\alpha$ outside $\xi(\theta_0)$, where $\theta_0$ can be taken to be any fixed positive number, independent of $P$. Obviously it pays to take $\theta_0$ small.

With this in mind, we define the *major arcs* $\mathfrak{M}$ to consist of the set $\xi(\theta_0)$, that is, the set of intervals (15.2) with $\theta$ replaced by $\theta_0$, and we define the minor arcs $\mathfrak{m}$ to consist of the complement of this set with respect to the interval $0 < \alpha < 1$.

**Lemma 15.1.** *Provided $n \geq 17$, we have*

$$\int_{\mathfrak{m}} |S(\alpha)|\, d\alpha \ll P^{n-3-\delta} \quad (\delta > 0).$$

*Proof.* We choose a set of numbers

$$\theta_0 < \theta_1 < \cdots < \theta_h = \tfrac{3}{4} + \delta.$$

Every real $\alpha$ lies in the set $\xi(\theta_h)$, because we can always find $a$, $q$ such that

$$q \leq P^{3/2}, \quad |q\alpha - a| < P^{-3/2},$$

and this implies that $\alpha$ is in $\xi(\frac{3}{4}+\delta)$. The minor arcs consist of the complement of $\xi(\theta_0)$, and we can regard this as the union of

$$\xi(\theta_h)-\xi(\theta_{h-1}), \quad \xi(\theta_{h-1})-\xi(\theta_{h-2}), \quad \cdots, \quad \xi(\theta_1)-\xi(\theta_0),$$

where the difference is meant in the sense of set theory.

In the set $\xi(\theta_g)-\xi(\theta_{g-1})$, the inequality (15.1) applies with $\theta=\theta_{g-1}$, so that

$$|S(\alpha)| \ll P^{n-\frac{1}{4}n\theta_{g-1}+\varepsilon}.$$

The measure of this set does not exceed the measure of $\xi(\theta_g)$, which is

$$\begin{aligned} &\ll \sum_{q\le P^{2\theta_g}} \sum_{a=1}^{q} q^{-1}P^{-3+2\theta_g} \\ &\ll P^{-3+4\theta_g}. \end{aligned}$$

Hence the contribution of this set to the integral of $|S(\alpha)|$ is

$$\ll P^{n-\frac{1}{4}n\theta_{g-1}-3+4\theta_g+\varepsilon}.$$

This is $\ll P^{n-3-\delta}$, since $n\ge 17$, provided

$$\theta_{g-1} > \frac{16}{17}\theta_g + \frac{4}{17}(\delta+\varepsilon).$$

We can choose $\theta_1,\ldots,\theta_h$ so near together that this is the case. Hence the result. □

Now we have to deal with the major arcs $\mathfrak{M}_{a,q}$, given by (15.2) with $\theta=\theta_0$. We put $2\theta_0=\Delta$, so that (15.2) becomes

$$(a,q)=1, \quad 1\le q\ll P^{\Delta}, \quad |q\alpha-a|\le P^{-3+\Delta}.$$

It will be convenient to enlarge the major arcs slightly; we replace them by the intervals $\mathfrak{M}'_{a,q}$ in which the last inequality is divided by $q$ on the left but not on the right:

$$(a,q)=1, \quad 1\le q\ll P^{\Delta}, \quad |\alpha-a/q|\le P^{-3+\Delta}. \tag{15.3}$$

This is plainly permissible, for the contribution made by the additional set is a part of the contribution made by the minor arcs to $\int|S(\alpha)|\,d\alpha$, and is therefore covered by the estimate of Lemma 15.1.

The object of this enlargement is to make the length of the intervals independent of $q$, as in Chapter 4. It is only possible to do this when $q$ is bounded by a small power of $P$, as here; but when it is possible, it leads to a slight simplification, in that the separation of the singular series from the singular integral takes place at an earlier point in the proof than it would do otherwise.

We define

$$S_{a,q} = \sum_{\mathbf{z} \pmod q} e\left(\frac{a}{q}C(\mathbf{z})\right),$$

$$I(\beta) = \int_{P\mathfrak{B}} e\left(\beta C(\boldsymbol{\xi})\right)\, d\boldsymbol{\xi},$$

where $P\mathfrak{B}$ is the box used in the earlier definition of $S(\alpha)$ in Chapter 13.

**Lemma 15.2.** *For $\alpha$ in an interval $\mathfrak{M}'_{a,q}$ we have, on putting $\alpha = \beta + a/q$,*

$$S(\alpha) = q^{-n} S_{a,q} I(\beta) + O(P^{n-1+2\Delta}).$$

*Proof.* The crude argument used in the proof of Lemma 4.2 suffices, the reason being that here (as there) $q$ is very small compared with $P$. We must now work in $n$ dimensions instead of in one dimension, however, and this means replacing a sum by an integral. We must make an allowance for the discrepancy between the number of integer points in a large box and the volume of the box.

Putting $\mathbf{x} = q\mathbf{y} + \mathbf{z}$, where $0 \le z_j < q$, we have to estimate the difference between

$$\sum_{\mathbf{y}} e\left(\beta C(q\mathbf{y} + \mathbf{z})\right) \quad \text{and} \quad \int e\left(\beta C(q\boldsymbol{\eta} + \mathbf{z})\right)\, d\boldsymbol{\eta}$$

where the conditions of summation on $\mathbf{y}$ are such as will make $0 < qy_j + z_j < P$, and similarly for $\boldsymbol{\eta}$. Thus the edges of the box for $\mathbf{y}$ are $\ll P/q$, and the allowance mentioned above is $(P/q)^{n-1}$. We have also to allow for the variation of the integrand in a box of edge-length 1. We have

$$\left| \frac{\partial}{\partial \eta_j} \beta C(q\boldsymbol{\eta} + \mathbf{z}) \right| \ll |\beta| q P^2 \ll q P^{-1+\Delta}.$$

The resulting error is obtained by multiplying by the volume of the region of integration, which is $\ll (P/q)^n$.

It follows that the difference between the sum and the integral is

$$\ll (P/q)^{n-1} + qP^{-1+\Delta} P^n q^{-n} \ll P^{n-1+\Delta} q^{1-n}.$$

(If $n$ were equal to 1, this would be $P^{\Delta}$, corresponding to $P^{\delta}$ in the proof of Lemma 4.2.) Summing over the $q^n$ values of $\mathbf{z}$, the final error term is

$$\ll P^{n-1+\Delta} q \ll P^{n-1+2\Delta}.$$

□

**Lemma 15.3.** *For a fixed cubic form $C(\mathbf{x})$ which does not represent zero, we have*

$$|S_{a,q}| \ll q^{\frac{7}{8}n+\varepsilon}.$$

*Proof.* We can apply alternative B of Lemma 13.4 to the sum $S(\alpha)$ with $\alpha = a/q$ and with $P = q$, and the box $\mathfrak{B}$ as the box $0 \le x_j < 1$, so that the box of summation becomes $0 \le x_j < q$. Take $\theta = \frac{1}{2} - \varepsilon$. Then either

$$|S(\alpha)| < q^{n-\frac{1}{4}n\theta+\varepsilon} \ll q^{\frac{7}{8}n+\varepsilon},$$

or $\alpha$ has a rational approximation $a'/q'$ satisfying

$$1 \le q' \ll P^{1-2\varepsilon}, \quad |q'\alpha - a'| < P^{-2}.$$

But the latter is impossible when $\alpha = a/q$, for it would give then $q' < q$ and

$$|q'(a/q) - a'| < q^{-2},$$

whereas it is obvious that $|q'(a/q) - a'| \ge 1/q$. Hence the estimate for $S(\alpha) = S_{a,q}$ holds. □

**Lemma 15.4.** *If $\mathfrak{M}'$ denotes the totality of the enlarged major arcs $\mathfrak{M}'_{a,q}$ then*

$$\int_{\mathfrak{M}'} S(\alpha)\, d\alpha = P^{n-3}\mathfrak{S}(P^\Delta)J(P^\Delta) + O(P^{n-3-\delta})$$

*for some $\delta > 0$, where*

$$\mathfrak{S}(P^\Delta) = \sum_{q \ll P^\Delta} \sum_{\substack{a=1 \\ (a,q)=1}}^{q} q^{-n} S_{a,q},$$

$$J(P^\Delta) = \int_{|\gamma|<P^\Delta} \left( \int_{\mathfrak{B}} e(\gamma C(\boldsymbol{\xi}))\, d\boldsymbol{\xi} \right) d\gamma.$$

*Proof.* The error term in Lemma 15.2, when integrated over $|\beta| < P^{-3+\Delta}$ and summed over $a$ and then over $q$ gives a final error term

$$\ll \sum_{q \ll P^{\Delta}} \sum_{a} P^{-3+\Delta} P^{n-1+2\Delta} \ll P^{n-4+5\Delta}.$$

This is $\ll P^{n-3-\delta}$ provided $\Delta$ is sufficiently small.

The main term in Lemma 15.2 gives

$$\sum_{q \ll P^{\Delta}} \sum_{\substack{a=1 \\ (a,q)=1}}^{q} q^{-n} S_{a,q} \int_{|\beta|<P^{-3+\Delta}} I(\beta)\, d\beta.$$

The summation and integration are independent, and the summation gives $\mathfrak{S}(P^{\Delta})$. The integral becomes

$$P^{-3} \int_{|\gamma|<P^{\Delta}} I(P^{-3}\gamma)\, d\gamma,$$

and this is $P^{n-3} J(P^{\Delta})$, since

$$\begin{aligned} I(P^{-3}\gamma) &= \int_{P\mathfrak{B}} e\left(P^{-3}\gamma C(\boldsymbol{\xi})\right)\, d\boldsymbol{\xi} \\ &= P^{n} \int_{\mathfrak{B}} e\left(\gamma C(\boldsymbol{\xi})\right)\, d\boldsymbol{\xi}. \end{aligned}$$

Hence the result. □

# 16

## Cubic forms: the singular integral

The singular integral is the integral occurring in Lemma 15.4, namely

$$J(\mu) = \int_{-\mu}^{\mu} \left( \int_{\mathcal{B}} e(\gamma C(\boldsymbol{\xi})) d\boldsymbol{\xi} \right) d\gamma, \tag{16.1}$$

where we have written $\mu$ for $P^{\Delta}$. It depends upon the box $\mathcal{B}$ which was used in the definition of the exponential sum $S(\alpha)$ in Chapter 13, and our object in the present chapter is to prove that we can choose $\mathcal{B}$ in such a way as to ensure that

$$J(\mu) \to J_0 > 0, \quad \text{as } \mu \to \infty. \tag{16.2}$$

We shall choose the box $\mathcal{B}$ so that it has for its centre a real solution $\xi_1^*, \ldots, \xi_n^*$ of the equation

$$C(\xi_1^*, \ldots, \xi_n^*) = 0. \tag{16.3}$$

This is a natural way to proceed, for our object is to obtain an asymptotic formula for $\mathcal{N}(P)$ which will show that $\mathcal{N}(P) \to \infty$; assuming that alternative A is excluded. If there were no real solution of $C(\boldsymbol{\xi}) = 0$ in $\mathcal{B}$, there would be none in $P\mathcal{B}$, and *a fortiori* $\mathcal{N}(P)$ would be 0.

We shall, in fact, take $\xi_1^*, \ldots, \xi_n^*$ to be rather more than an arbitrary real solution of $C(\boldsymbol{\xi}) = 0$; we shall take it to be a non-singular solution[1], in which none of $\xi_1^*, \ldots, \xi_n^*$ is 0. This is a convenient choice to ensure the truth of (16.2), and may even be essential.

The existence of such a solution is easily proved. For any real $\xi_2, \ldots, \xi_n$ we can find a real $\xi_1$ to satisfy $C(\xi_1, \ldots, \xi_n) = 0$, and it suffices to ensure that

$$\xi_1 \neq 0 \quad \text{and} \quad \partial C / \partial \xi_1 \neq 0.$$

[1] That is, one for which the partial derivatives of $C$ are not all 0.

For any $\xi_2, \ldots, \xi_n$, the equation for $\xi_1$ is of the form

$$c_{111}\xi_1^3 + F\xi_1^2 + G\xi_1 + H = 0,$$

where $F, G, H$ are forms in $\xi_2, \ldots, \xi_n$ of degrees $1, 2, 3$ respectively. Provided $H \neq 0$ (and we note that $H$ cannot vanish identically) we shall have $\xi_1 \neq 0$. Let $D(\xi_2, \ldots, \xi_n)$ be the discriminant of the cubic equation in $\xi_1$. Then, provided $D \neq 0$, we shall have $\partial C/\partial \xi_1 \neq 0$. We can suppose that $D$ does not vanish identically, for then the double root for $\xi_1$ is determined rationally by $\xi_2, \ldots, \xi_n$ and we get rational solutions of $C(\boldsymbol{\xi}) = 0$.

Thus we can find the desired non-singular real solution $\xi_1^*, \ldots, \xi_n^*$ of (16.3). We take $\mathcal{B}$ to be a small cube around this point, say

$$|\xi_j - \xi_j^*| < \rho, \quad (1 \leq j \leq n). \tag{16.4}$$

**Lemma 16.1.** *If $\rho$ is chosen sufficiently small, (16.2) holds.*

*Proof.* We have

$$\begin{aligned} J(\mu) &= \int_{-\mu}^{\mu} \left( \int_{\mathcal{B}} e(\gamma C(\boldsymbol{\xi}) d\boldsymbol{\xi} \right) d\gamma \\ &= \int_{\mathcal{B}} \frac{\sin 2\pi\mu C(\boldsymbol{\xi})}{\pi C(\boldsymbol{\xi})} d\boldsymbol{\xi} \\ &= \int_{-\rho}^{\rho} \cdots \int_{-\rho}^{\rho} \frac{\sin 2\pi\mu C(\boldsymbol{\xi}^* + \boldsymbol{\eta})}{\pi C(\boldsymbol{\xi}^* + \boldsymbol{\eta})} d\boldsymbol{\eta}, \end{aligned} \tag{16.5}$$

where $\boldsymbol{\xi} = \boldsymbol{\xi}^* + \boldsymbol{\eta}$.

For any $\boldsymbol{\eta}$, we have

$$C(\boldsymbol{\xi}^* + \boldsymbol{\eta}) = c_1\eta_1 + \cdots + c_n\eta_n + P_2(\boldsymbol{\eta}) + P_3(\boldsymbol{\eta}), \tag{16.6}$$

where $P_2(\boldsymbol{\eta})$, $P_3(\boldsymbol{\eta})$ are forms of degrees $2, 3$ in $\boldsymbol{\eta}$. We have

$$c_1 = \frac{\partial C}{\partial \xi_1}(\xi_1^*, \ldots, \xi_n^*) \neq 0.$$

Without loss of generality we can suppose $c_1 = 1$.

For $|\boldsymbol{\eta}| < \rho$, we have

$$|C(\boldsymbol{\xi}^* + \boldsymbol{\eta})| < \sigma,$$

where $\sigma = \sigma(\rho)$ is small with $\rho$. Put $C(\boldsymbol{\xi}^* + \boldsymbol{\eta}) = \zeta$. Then, if $\rho$ is sufficiently small, we can invert the relation (16.6) and express $\eta_1$ in terms of $\eta_2, \ldots, \eta_n$ by means of a power series. This will be one of the form

$$\eta_1 = \zeta - c_2\eta_2 - \cdots - c_n\eta_n + P(\zeta, \eta_2, \ldots, \eta_n),$$

where $P$ is a multiple power series beginning with terms of degree 2 at least. Hence

$$\frac{\partial \eta_1}{\partial \zeta} = 1 + P_1(\zeta, \eta_2, \ldots, \eta_n),$$

and by taking $\rho$ sufficiently small we can ensure that $|P_1| < 1/2$ for

$$|\eta_2| < \rho, \ldots, |\eta_n| < \rho, |\zeta| < \sigma.$$

Making a change of variable from $\eta_1$ to $\zeta$ in (16.5), we obtain

$$J(\mu) = \int_{-\sigma}^{\sigma} \frac{\sin 2\pi\mu\zeta}{\pi\zeta} V(\zeta) d\zeta, \tag{16.7}$$

where

$$V(\zeta) = \int_{\mathcal{B}'} \{1 + P_1(\zeta, \eta_2, \ldots, \eta_n)\} \, d\eta_2 \cdots d\eta_n,$$

in which $\mathcal{B}'$ denotes the part of the $(n-1)$-dimensional cube

$$|\eta_2| < \rho, \ldots, |\eta_n| < \rho$$

in which $|\eta_1| < \rho$, that is, in which

$$|\zeta - c_2\eta_2 - \cdots - c_n\eta_n + P(\zeta, \eta_2, \ldots, \eta_n)| < \rho.$$

It is clear that $V(\zeta)$ is a continuous function of $\zeta$ for $|\zeta|$ sufficiently small. It is also easily seen that $V(\zeta)$ is a function of bounded variation, since it has left and right derivatives at every value of $\zeta$, and these are bounded. Hence, by Fourier's integral theorem applied to (16.7), we have

$$\lim_{\mu \to \infty} V(\mu) = V(0).$$

Now $V(0)$ is a positive number, for the cube $\mathcal{B}'$ contains any sufficiently small $(n-1)$-dimensional cube centred at the origin, and in such a cube we have $1 + P_1 > 1/2$. This proves the result. □

# 17

# Cubic forms: the singular series

Putting together the results of Lemmas 15.1, 15.4 and 16.1, we have now proved that if $n \geq 17$, and if alternative A of Lemma 13.4 is excluded, and if the box $\mathfrak{B}$ is suitably chosen, then the number $\mathcal{N}(P)$ of integer points with $C(\mathbf{x}) = 0$ in the box $P\mathfrak{B}$ satisfies

$$\mathcal{N}(P) = P^{n-3}\mathfrak{S}(P^{\Delta})\{J_0 + o(1)\} + O(P^{n-3-\delta}),$$

where $\delta > 0$. The series for $\mathfrak{S}(P^{\Delta})$ in Lemma 15.4 converges absolutely if continued to infinity, provided $n \geq 17$ and alternative A is excluded, since then

$$q^{-n}|S_{a,q}| \ll q^{-\frac{1}{8}n+\varepsilon-2-\delta}$$

by Lemma 15.3. By the work of Chapter 14, the alternative A implies that $C(\mathbf{x}) = 0$ has a non-trivial integral solution. Thus we have proved:

**Theorem 17.1.** *If $n \geq 17$ and $\mathfrak{S} > 0$, then the equation $C(\mathbf{x}) = 0$ has a non-trivial integral solution.*

For if there is no non-trivial solution we get

$$\mathcal{N}(P) \sim P^{n-3}\mathfrak{S}J_0 \quad \text{as } P \to \infty,$$

whence $\mathcal{N}(P) \to \infty$, giving a contradiction.

Here, of course, $\mathfrak{S}$ denotes the singular series continued to infinity, i.e.

$$\mathfrak{S} = \sum_{q=1}^{\infty} \sum_{\substack{a=1 \\ (a,q)=1}}^{q} q^{-n}S_{a,q}.$$

It remains to prove that $\mathfrak{S} > 0$ for every cubic form in 17 or more variables. The proof of Lemma 5.1 applies to exponential sums in general (as remarked at the time), and shows that if

$$A(q) = \sum_{\substack{a=1 \\ (a,q)=1}}^{q} q^{-n} S_{a,q}$$

then $A(q)$ is a multiplicative function of $q$ (for relatively prime values of $q$). Provided $n \geq 17$ and $C(\mathbf{x})$ does not represent zero, we have

$$|A(q)| \ll q^{1-\frac{1}{8}n+\varepsilon} \ll q^{-1-\delta}$$

by Lemma 15.3. Hence $\mathfrak{S} = \sum_q A(q)$ is absolutely convergent, and it follows that

$$\mathfrak{S} = \prod_p \chi(p),$$

where

$$\chi(p) = 1 + \sum_{\nu=1}^{\infty} A(p^\nu).$$

We also have (under the above conditions)

$$|\chi(p) - 1| \ll p^{-1-\delta},$$

so there exists $p_0$ such that

$$\prod_{p>p_0} \chi(p) \geq \frac{1}{2},$$

as in the Corollary to Lemma 5.2. The argument of Lemma 5.3 shows that

$$\chi(p) = \lim_{\nu\to\infty} \frac{M(p^\nu)}{p^{\nu(n-1)}},$$

where $M(p^\nu)$ denotes the number of solutions of

$$C(x_1, \ldots, x_n) \equiv 0 \pmod{p^\nu}, \quad 0 \leq x_j < p^\nu.$$

Hence, in order to prove that $\mathfrak{S} > 0$ (under the present conditions) it will suffice to prove that, for each prime $p$, we have

$$M(p^\nu) \geq C_p p^{\nu(n-1)}, \quad C_p > 0, \tag{17.1}$$

for all sufficiently large $\nu$.

When we were dealing with forms of additive type, we found that the existence of just one solution of the congruence

$$a_1 x_1^k + \cdots + a_n x_n^k \equiv 0 \pmod{p^{\gamma_1}},$$

with not all of $x_1, \ldots, x_n$ divisible by $p$, implied the truth of (17.1) for all sufficiently large $\nu$, provided $\gamma_1$ was a suitable exponent depending on $p$ and $k$ and on the powers of $p$ dividing the various coefficients $a_j$. For a general form, the position is not quite so simple. It appears that one needs more than just a solution (mod $p^{\gamma_1}$); one needs a solution for which the partial derivatives $\partial C/\partial x_1, \ldots, \partial C/\partial x_n$ are not all divisible by too high a power of $p$. For an additive form, there is an obvious limit to the power of $p$, since the derivative with respect to $x_j$ is $a_j k x_j^{k-1}$, and there is some $j$ for which $x_j$ is not divisible by $p$.

**Definition.** Let $p$ be a prime and $\ell$ a positive integer. We say $C(\mathbf{x})$ has the property $\mathcal{A}(p^\ell)$ if there is a solution of

$$C(x_1, \ldots, x_n) \equiv 0 \pmod{p^{2\ell-1}} \tag{17.2}$$

with

$$\partial C/\partial x_i \equiv 0 \pmod{p^{\ell-1}} \quad \text{for all } i \tag{17.3}$$

and

$$\partial C/\partial x_i \not\equiv 0 \pmod{p^{\ell}} \quad \text{for some } i. \tag{17.4}$$

**Lemma 17.1.** *Suppose $C(\mathbf{x})$ has the property $\mathcal{A}(p^\ell)$. Then*

$$M\left(p^{2\ell-1+\nu}\right) \geq p^{(n-1)\nu},$$

*and consequently $\chi(p) > 0$.*

*Proof.* We prove by induction on $\nu$ that the congruence

$$C(x_1, \ldots, x_n) \equiv 0 \pmod{p^{2\ell-1+\nu}} \tag{17.5}$$

has at least $p^{(n-1)\nu}$ solutions satisfying (17.3) and (17.4), these solutions being mutually incongruent to the modulus $p^{\ell+\nu}$ (and therefore *a fortiori* to the modulus $p^{2\ell-1+\nu}$). This will imply the result. When $\nu = 0$ the assertion is simply that of the hypothesis.

For any integers $x_1, \ldots, x_n, u_1, \ldots, u_n$ we have

$$C(\mathbf{x} + p^{\ell+\nu}\mathbf{u}) \equiv C(\mathbf{x}) + p^{\ell+\nu}(u_1 \partial C/\partial x_1 + \cdots) \pmod{p^{2\ell+2\nu}}.$$

We assume that the result just stated holds for a particular $\nu$, and we take $x_1, \ldots, x_n$ to be any one of the $p^{(n-1)\nu}$ solutions of (17.5) which satisfy (17.3) and (17.4), these solutions being mutually incongruent to the modulus $p^{\ell+\nu}$. We can put

$$C(\mathbf{x}) = a p^{2\ell-1+\nu}, \quad \partial C/\partial x_i = D_i p^{\ell-1},$$

where $a, D_1, \ldots, D_n$ are integers and $D_i \not\equiv 0 \pmod{p}$ for some $i$. The congruence

$$C(\mathbf{x} + p^{\ell+\nu}\mathbf{u}) \equiv 0 \pmod{p^{2\ell+\nu}}$$

holds if

$$a + D_1 u_1 + \cdots + D_n u_n \equiv 0 \pmod{p}.$$

This has $p^{n-1}$ solutions in $\mathbf{u}$, mutually incongruent $(\mathrm{mod}\ p)$.

Thus corresponding to each $\mathbf{x}$ we get $p^{n-1}$ values of $\mathbf{y} = \mathbf{x} + p^{\ell+\nu}\mathbf{u}$. These satisfy

$$C(\mathbf{y}) \equiv 0 \pmod{p^{2\ell+\nu}}, \quad \mathbf{y} \equiv \mathbf{x} \pmod{p^{\ell+\nu}}.$$

From the latter it follows that each $\mathbf{y}$ satisfies (17.3) and (17.4). We obtain altogether $p^{(n-1)(\nu+1)}$ values for $\mathbf{y}$ and they are mutually incongruent to the modulus $p^{\ell+\nu+1}$. Hence the assertion holds for $\nu + 1$ in place of $\nu$, and this proves the result. □

The proof that for each $p$ there is some $\ell$ such that $C(\mathbf{x})$ has the property $\mathcal{A}(p^{\ell})$ forms the subject of the next chapter.

# 18

# Cubic forms: the $p$-adic problem

The assertion that for any $p$ there exists some $\ell$ such that $C(\mathbf{x})$ has the property $\mathcal{A}(p^{\ell})$ is equivalent to the assertion that the equation $C(\mathbf{x}) = 0$ has a non-singular solution in the $p$-adic number field. We shall prove that this is true for $n \geq 10$. Several mathematicians have proved independently that the equation $C(\mathbf{x}) = 0$ always has a non-trivial $p$-adic solution provided $n \geq 10$, and any one of their proofs would serve our purpose, because (as Professor Lewis pointed out to me) it is possible to deduce a non-singular solution from any non-trivial solution. However, I prefer to follow my own proof, as this was designed to lead directly to the property $\mathcal{A}(p^{\ell})$.

Let $N = \frac{1}{2}n(n+1)$. Let $\mathcal{C}$ denote the matrix of $n$ rows and $N$ columns whose general element is $c_{ijk}$, where $i$ indicates the row and the pair $j, k$ (with $j \leq k$) indicates the column, on the understanding that these pairs are arranged in some fixed order. Let $\Delta$ denote a typical determinant of order $n$ formed from any $n$ columns of $\mathcal{C}$, the number of possible determinants being $\binom{N}{n}$. We can assume (multiplying $C(\mathbf{x})$ by a factor of 6 if necessary) that the $c_{ijk}$ are integral, hence that the various $\Delta$ are integral.

**Definition.** Let $h(\mathcal{C})$ denote the highest common factor of all the determinants $\Delta$, if they are not all 0, and in the latter case let $h(\mathcal{C}) = 0$.

**Lemma 18.1.** *Let*

$$x_i' = \sum_{r=1}^{n} q_{ir} x_r, \quad (1 \leq i \leq n),$$

*be a linear transformation with integral coefficients $q_{ir}$ of determinant*

*$q \neq 0$, and let*

$$C(x_1, \ldots, x_n) = C'(x'_1, \ldots, x'_n)$$

*identically. Then $h(C)$ is divisible by $qh(C')$.*

*Proof.* The coefficients in the two forms $C$ and $C'$ are related by

$$c_{rst} = \sum_{i=1}^{n} \sum_{j=1}^{n} \sum_{k=1}^{n} q_{ir} q_{js} q_{kt} c'_{ijk}.$$

In defining the matrix $\mathcal{C}$ above, we chose a one-to-one correspondence between pairs $j, k$ with $1 \leq j \leq k \leq n$ and integers $1 \leq \mu \leq N$. Thus the general element of $\mathcal{C}'$ is $c'_{ijk} = c'_{i\mu}$, say, where $i = 1, 2, \ldots, n$ and $\mu = 1, 2, \ldots, N$. Similarly, representing the pair $s, t$ with $s \leq t$ by $\nu$, the general element of $\mathcal{C}$ is $c_{rst} = c_{r\nu}$. Put

$$u_{\mu\nu} = \begin{cases} q_{js} q_{kt}, & j = k, \\ q_{js} q_{kt} + q_{ks} q_{jt}, & j < k, \end{cases}$$

where $\mu$ denotes the pair $j, k$ and $\nu$ the pair $s, t$. Then the relation between the two sets of coefficients can be written

$$c_{r\nu} = \sum_{i=1}^{n} q_{ir} \sum_{\mu=1}^{N} c'_{i\mu} u_{\mu\nu},$$

where $r = 1, 2, \ldots, n$ and $\nu = 1, 2, \ldots, N$. In matrix notation this is

$$\mathcal{C} = \mathcal{Q}^T \mathcal{C}' \mathcal{U},$$

where $\mathcal{Q} = (q_{ir})$ is an $n \times n$ matrix and $\mathcal{U} = (u_{\mu\nu})$ is an $N \times N$ matrix, and $T$ denotes the transpose.

Let $\Delta$ be the determinant formed from the columns $\nu_1, \ldots, \nu_n$ of $\mathcal{C}$, or symbolically:

$$\Delta = (\det \mathcal{C})^{1,\ldots,n}_{\nu_1,\ldots,\nu_n}.$$

Since the matrix $\mathcal{Q}$ has determinant $q$, it follows that

$$\pm\Delta = q\,(\det \mathcal{C}'\mathcal{U})^{1,\ldots,n}_{\nu_1,\ldots,\nu_n}.$$

By a well known result we have

$$(\det \mathcal{C}'\mathcal{U})^{1,\ldots,n}_{\nu_1,\ldots,\nu_n} = \sum_{\rho_1,\ldots,\rho_n} (\det \mathcal{C}')^{1,\ldots,n}_{\rho_1,\ldots,\rho_n} (\det \mathcal{U})^{\rho_1,\ldots,\rho_n}_{\nu_1,\ldots,\nu_n},$$

where the summation is over all $\binom{N}{n}$ selections of $\rho_1, \ldots, \rho_n$ from $1, \ldots, N$ without regard to order.

In each term of the sum, the first factor is one of the determinants $\Delta'$ of order $n$ that can be formed from $\mathcal{C}'$, and the second factor is an integer. Hence the sum is divisible by $h(C')$ and it follows that $\Delta$ is divisible by $qh(C')$. This proves the result. □

**Corollary.** *$h(C)$ is an arithmetic invariant of $C$. That is, it has the same value for any two equivalent forms.*

*Proof.* If $C$ and $C'$ are equivalent forms the lemma applies with $q = 1$, and shows that $h(C)$ is divisible by $h(C')$. Similarly $h(C')$ is divisible by $h(C)$, whence the result. □

**Lemma 18.2.** *If $C(\mathbf{x})$ is non-degenerate then $h(C) \neq 0$.*

*Proof.* If $h(C) = 0$ then all the determinants of order $n$ formed from $\mathcal{C}$ vanish, that is the $n$ rows of $\mathcal{C}$ are linearly dependent. Thus there exist $p_1, \ldots, p_n$, not all 0, such that

$$\sum_{i=1}^{n} p_i c_{ijk} = 0$$

for all $j, k$; and we can take $p_1, \ldots, p_n$ to be integers with highest common factor 1. Since

$$\frac{1}{3}\frac{\partial C}{\partial x_i} = \sum_j \sum_k c_{ijk} x_j x_k,$$

we have

$$p_1 \frac{\partial C}{\partial x_1} + \cdots + p_n \frac{\partial C}{\partial x_n} = 0$$

identically in $x_1, \ldots, x_n$. It is well known that there exists an $n \times n$ matrix of integers $p_{ir}$, of determinant $\pm 1$, such that $p_{in} = p_i$ for $i = 1, 2, \ldots, n$. Putting

$$x_i = \sum_{r=1}^{n} p_{ir} y_r,$$

we have

$$\frac{\partial C}{\partial y_n} = 0$$

identically, and consequently $C(\mathbf{x})$ is equivalent to a form in $y_1, \ldots, y_{n-1}$ and is degenerate. □

The converse of this last lemma is also true, for the above argument is reversible, but it will not be needed.

**Lemma 18.3.** *If $n \geq 4$ and $C(\mathbf{x})$ does not have the property $\mathcal{A}(p)$, then $C(\mathbf{x})$ is equivalent to a form of the type*

$$C'(x_1, x_2, x_3) + pC''(x_1, \ldots, x_n). \tag{18.1}$$

*Proof.* By a theorem of Chevalley[1] there is a solution of $C(\mathbf{x}) \equiv 0 \pmod{p}$ other than $\mathbf{x} = \mathbf{0}$, since the number of variables exceeds the degree of the congruence. As $C(\mathbf{x})$ does not have the property $\mathcal{A}(p)$, we must have

$$\frac{\partial C}{\partial x_i} \equiv 0 \pmod{p}$$

for all $i$. After a suitable integral unimodular transformation, we can take the solution in question to be

$$x_1 = 1, \quad x_2 = x_3 = \cdots = x_n = 0.$$

Then $C(\mathbf{x})$ has the form

$$apx_1^3 + px_1^2(b_2x_2 + \cdots + b_nx_n) + x_1B(x_2, \ldots, x_n) + C_{n-1}(x_2, \ldots, x_n),$$

where $B$ and $C_{n-1}$ are quadratic and cubic forms respectively. Indeed the coefficients of $x_1^2x_j$ are all divisible by $p$ because they are the values of

$$\frac{\partial C}{\partial x_2}, \ldots, \frac{\partial C}{\partial x_n}$$

at the solution.

If some of the coefficients of $B$ are not divisible by $p$, we can choose $x_2, \ldots, x_n$ so that

$$B(x_2, \ldots, x_n) \not\equiv 0 \pmod{p},$$

by taking values of the type $1, 0, \ldots, 0$ or of the type $1, 1, 0, \ldots, 0$. We can then choose $x_1$ so that

$$x_1B(x_2, \ldots, x_n) + C_{n-1}(x_2, \ldots, x_n) \equiv 0 \pmod{p}$$

and this gives a solution of $C(\mathbf{x}) \equiv 0 \pmod{p}$ with $\partial C/\partial x_1 \not\equiv 0 \pmod{p}$, contrary to the hypothesis.

[1] See [23], for example.

We can therefore assume that all the coefficients of $B$ are divisible by $p$. Thus $C(\mathbf{x})$ is equivalent to

$$px_1Q_n(x_1, x_2, \ldots, x_n) + C_{n-1}(x_2, \ldots, x_n),$$

where $Q_n$ is a quadratic form.

If $n \geq 5$, we can put $x_1 = 0$ and apply the argument to $C_{n-1}(x_2, \ldots, x_n)$ since this form also cannot have the property $\mathcal{A}(p)$. Thus $C_{n-1}$ is equivalent to

$$px_2Q_{n-1}(x_2, \ldots, x_n) + C_{n-2}(x_3, \ldots, x_n).$$

This process continues until we reach $C_3(x_{n-2}, x_{n-1}, x_n)$, to which Chevalley's theorem does not apply. Hence $C(\mathbf{x})$ is equivalent to a form of the type

$$p(x_1Q_n + \cdots + x_{n-3}Q_4) + C_3(x_{n-2}, x_{n-1}, x_n).$$

Reversing the order of writing the variables, we obtain a form of the type (18.1). □

**Lemma 18.4.** *If, in the result of Lemma 18.3, the form*

$$C''(0, 0, 0, x_4, \ldots, x_n)$$

*in $x_4, \ldots, x_n$ has the property $\mathcal{A}(p^\lambda)$, then $C(\mathbf{x})$ has the property $\mathcal{A}(p^\ell)$ for some $\ell \leq \lambda + 1$.*

*Proof.* In the proof of Lemma 17.1 we saw that if a form $C^*$ had property $\mathcal{A}(p^\lambda)$ then for every $\nu \geq 0$ the congruences

$$C^*(\mathbf{x}) \equiv 0 \pmod{p^{2\lambda-1+\nu}} \tag{18.2}$$

$$\frac{\partial C^*}{\partial x_i} \equiv 0 \pmod{p^{\lambda-1}}, \tag{18.3}$$

were soluble for all $i$, and in addition, for some $j$,

$$\frac{\partial C^*}{\partial x_j} \not\equiv 0 \pmod{p^\lambda}. \tag{18.4}$$

For brevity, we express (18.3) and (18.4) by

$$p^{\lambda-1} \| \left( \frac{\partial C^*}{\partial x_1}, \ldots, \frac{\partial C^*}{\partial x_n} \right).$$

The hypothesis that $C''(0, 0, 0, x_4, \ldots, x_n)$ has property $\mathcal{A}(p^\lambda)$ implies (taking $\nu = 1$) the existence of integers $x_4, \dots, x_n$ such that

$$C''(0,0,0,x_4,\dots,x_n) \equiv 0 \pmod{p^{2\lambda}}, \quad p^\lambda \| \left( \frac{\partial C''}{\partial x_4}, \dots, \frac{\partial C''}{\partial x_n} \right).$$

Hence

$$C(0,0,0,x_4,\dots,x_n) \equiv 0 \pmod{p^{2\lambda+1}}, \quad p^\lambda \| \left( \frac{\partial C}{\partial x_4}, \dots, \frac{\partial C}{\partial x_n} \right).$$

If, for these values of $x_1, x_2, \dots, x_n$, we define $\ell$ by

$$p^{\ell-1} \| \left( \frac{\partial C}{\partial x_1}, \dots, \frac{\partial C}{\partial x_n} \right),$$

then $\ell \le \lambda + 1$ and $C(\mathbf{x})$ has property $\mathcal{A}(p^\ell)$. □

**Lemma 18.5.** *If $n \ge 10$ and $C(\mathbf{x})$ does not have any of the properties $\mathcal{A}(p)$, $\mathcal{A}(p^2)$, $\mathcal{A}(p^3)$, then it is equivalent to a form of the type*

$$C^*(x_1, x_2, \dots, x_9, px_{10}, \dots, px_n). \tag{18.5}$$

*Proof.* In the expression (18.1) for a form equivalent to $C$, which we denote again by $C$, we put $x_i = py_i$ for $i = 1, 2, 3$. This gives

$$\begin{aligned} &C(py_1, py_2, py_3, x_4, \dots, x_n) \\ &\quad \equiv p^3 C'(y_1, y_2, y_3) + pC''(py_1, py_2, py_3, x_4, \dots, x_n). \end{aligned}$$

Ignoring multiples of $p^3$, we have

$$\begin{aligned} &C(py_1, py_2, py_3, x_4, \dots, x_n) \\ &\quad \equiv p^2 C_{1,2}(y_1, y_2, y_3 | x_4, \dots, x_n) + pC''(0,0,0,x_4,\dots,x_n) \pmod{p^3}, \end{aligned} \tag{18.6}$$

where $C_{1,2}$ denotes a form which is of first degree in $y_1, y_2, y_3$ and of second degree in $x_4, \dots, x_n$.

By Lemma 18.4, the form $C''(0,0,0,x_4,\dots,x_n)$ does not have either property $\mathcal{A}(p)$ or $\mathcal{A}(p^2)$. We apply Lemma 18.3 to this form and put $x_i = py_i$ for $i = 4, 5, 6$ in the result. Neglecting multiples of $p^2$, we obtain

$$\begin{aligned} &C''(0,0,0,py_4,py_5,py_6,x_7,\dots,x_n) \\ &\quad \equiv pC^{(3)}(0,\dots,0,x_7,\dots,x_n) \pmod{p^2}. \end{aligned} \tag{18.7}$$

Further, by Lemma 18.4, the form $C^{(3)}(0,\dots,0,x_7,\dots,x_n)$ in $x_7, \dots, x_n$ does not have the property $\mathcal{A}(p)$.

Putting $x_i = py_i$ for $i = 4, 5, 6$ in (18.6), and using (18.7), we obtain a result which can be written

$$C(py_1, \ldots, py_6, x_7, \ldots, x_n)$$
$$\equiv p^2(y_1Q_1 + y_2Q_2 + y_3Q_3) + p^2C^{(3)}(0, \ldots, 0, x_7, \ldots, x_n) \pmod{p^3}, \tag{18.8}$$

where $Q_1, Q_2, Q_3$ are quadratic forms in $x_7, \ldots, x_n$. It should be noted that $y_4, y_5, y_6$ do not appear on the right of (18.8).

Suppose one of the quadratic forms, say $Q_1$, is not identically $\equiv 0 \pmod{p}$. Then there exists $x_7, \ldots, x_n$ for which $Q_1 \not\equiv 0 \pmod{p}$, and we can choose $y_1, y_2, y_3$ so that

$$y_1Q_1 + y_2Q_2 + y_3Q_3 + C^{(3)}(0, \ldots, 0, x_7, \ldots, x_n) \equiv 0 \pmod{p}.$$

This gives

$$C(py_1, \ldots, py_6, x_7, \ldots, x_n) \equiv 0 \pmod{p^3},$$

the values of $y_4, y_5, y_6$ being arbitrary. Also

$$\frac{\partial C}{\partial y_1}(py_1, \ldots, py_6, x_7, \ldots, x_n) \equiv p^2Q_1 \not\equiv 0 \pmod{p^3}.$$

Taking $x_i = py_i$ for $i = 1, \ldots, 6$ and noting that $\partial/\partial x_1 = p^{-1}\partial/\partial y_1$,we have values of $x_1, \ldots, x_n$ for which $C \equiv 0 \pmod{p^3}$ and $\partial C/\partial x_1 \not\equiv 0 \pmod{p^2}$. This contradicts the hypothesis that $C$ does not have either of the properties $\mathcal{A}(p)$ or $\mathcal{A}(p^2)$.

Thus $Q_1, Q_2, Q_3$ are all identically $\equiv 0 \pmod{p}$ and (18.8) becomes

$$C(py_1, \ldots, py_6, x_7, \ldots, x_n) \equiv p^2C^{(3)}(0, \ldots, 0, x_7, \ldots, x_n) \pmod{p^3}.$$

Finally, we apply Lemma 18.3 to the form $C^{(3)}(0, \ldots, 0, x_7, \ldots, x_n)$ which (as already noted) does not have the property $\mathcal{A}(p)$. We obtain

$$C^{(3)}(0, \ldots, 0, x_7, \ldots, x_n) \equiv C^{(4)}(x_7, x_8, x_9) \pmod{p}.$$

Putting $x_i = py_i$ for $i = 1, \ldots, 9$ we get

$$C(py_1, \ldots, py_9, x_{10}, \ldots, x_n) \equiv 0 \pmod{p^3},$$

and this holds identically in $y_1, \ldots, y_9, x_{10}, \ldots, x_n$. Denoting the form on the left by

$$p^3C^*(y_1, \ldots, y_9, x_{10}, \ldots, x_n),$$

we have the identity

$$C(x_1, \ldots, x_n) = C^*(x_1, \ldots, x_9, px_{10}, \ldots, px_n).$$

□

**Lemma 18.6.** *Suppose $n \geq 10$. If in the result of Lemma 18.5 the form $C^*(x_1, \ldots, x_n)$ has the property $\mathcal{A}(p^\lambda)$ then the form $C(x_1, \ldots, x_n)$ has the property $\mathcal{A}(p^\ell)$ for some $\ell \leq \lambda + 3$.*

*Proof.* As we observed at the start of the proof of Lemma 18.4, the hypothesis that $C^*$ has the property $\mathcal{A}(p^\lambda)$ implies the existence of values $y_1, \ldots, y_n$ such that

$$C^*(y_1, \ldots, y_n) \equiv 0 \pmod{p^{2\lambda+2}}, \quad p^{\lambda-1} \| \left( \frac{\partial C^*}{\partial y_1}, \ldots, \frac{\partial C^*}{\partial y_n} \right).$$

Since

$$C(py_1, \ldots, py_9, y_{10}, \ldots, y_n) = p^3 C^*(y_1, \ldots, y_9, y_{10}, \ldots, y_n)$$

identically, we have

$$C(py_1, \ldots, py_9, y_{10}, \ldots, y_n) \equiv 0 \pmod{p^{2\lambda+5}}$$

and one, at least, of $\partial C/\partial y_1, \ldots, \partial C/\partial y_9, \partial C/\partial y_{10}, \ldots, \partial C/\partial y_n$ is not divisible by $p^{\lambda+3}$. Putting $x_i = py_i$ for $i = 1, \ldots, 9$ and $x_i = y_i$ for $i \geq 10$, we have one at least of $\partial C/\partial x_1, \ldots, \partial C/\partial x_n$ not divisible by $p^{\lambda+3}$, whence the result. □

**Lemma 18.7.** *Any non-degenerate cubic form with integral coefficients in at least 10 variables has the property $\mathcal{A}(p^\ell)$ for every prime $p$ and a suitable $\ell$ depending on $p$. There is an upper bound for $\ell$ depending only on the cubic form.*

*Proof.* Suppose $C(\mathbf{x})$ is a cubic form with integer coefficients which does not have any of the properties $\mathcal{A}(p), \mathcal{A}(p^2), \ldots, \mathcal{A}(p^{3m})$, where $m$ is a positive integer. By Lemma 18.5, this form is equivalent to a form of type (18.5). This implies that there is a linear transformation

$$x_i' = \sum_{r=1}^{n} q_{ir} x_r, \quad (1 \leq i \leq n),$$

with integral coefficients and determinant $p^{n-9}$, which transforms $C(x_1, \ldots, x_n)$ into another form $C^{(1)}(x_1', \ldots, x_n')$ with integral coefficients. By Lemma 18.6, the form $C^{(1)}$ does not have any of the properties $\mathcal{A}(p)$, $\mathcal{A}(p^2), \ldots, \mathcal{A}(p^{3m-3})$. By repetition, it follows that there is a linear transformation with integral coefficients and determinant $p^{(n-9)m}$ which transforms $C(\mathbf{x})$ into a form $C^{(m)}(\mathbf{y})$ with integral coefficients.

It follows from Lemma 18.1 that $h(C)$ is divisible by $p^{(n-9)m}$. Further $h(C)$ is a positive integer by Lemma 18.2. Thus

$$(n-9)m \leq \log h(C)/\log p \leq \log h(C)/\log 2,$$

and this gives an upper bound for $m$, independent of $p$. This completes the proof of Lemma 18.7. □

In view of Theorem 17.1 and the subsequent remarks of Chapter 17, we see that Lemma 18.7 completes the proof of the following result.

**Theorem 18.1.** *If $C(x_1, \ldots, x_n)$ is any cubic form with integral coefficients, and $n \geq 17$, the equation*

$$C(x_1, \ldots, x_n) = 0$$

*has a solution in integers $x_1, \ldots, x_n$, not all 0.*

# 19

# Homogeneous equations of higher degree

In [6], Birch has given a far-reaching extension of the method by which we have treated the homogeneous cubic equation, but this involves some important modifications. He considers the problem of solving a homogeneous equation, or a system of simultaneous homogeneous equations (all of the same degree). Here one is faced by two serious difficulties. In the first place, even for a single equation of degree $k > 3$, we do not in general know of any reasonable function $n_0$ of $k$ which will be such that, if $n \geq n_0(k)$, the congruence conditions corresponding to the equation will be satisfied for every prime. (We know there is *some* function of $k$ for which the equation is soluble in the $p$-adic field, by the work of Brauer, quoted in Chapter 11, but this leads to an astronomical value. Moreover, a solution in the $p$-adic field would not be quite enough; we need a non-singular solution in order to be sure that we can satisfy the congruence conditions.) Thus we must postulate that the congruence conditions are satisfied for each prime $p$. We must also postulate that the equation, or system of equations, is soluble in the real number-field, with a non-singular solution.

Secondly — and this is more important — even these postulates are not always enough to ensure the solubility of the equation in integers (or rational numbers). The following example was shown to me by Swinnerton-Dyer;

$$3(x_1^2 + \cdots + x_r^2)^3 + 4(x_{r+1}^2 + \cdots + x_s^2)^3 = 5(x_{s+1}^2 + \cdots + x_n^2)^3 \quad (19.1)$$

where $r < s < n$. It is known from the work of Selmer [78] that the equation

$$3X^3 + 4Y^3 = 5Y^3$$

is insoluble except with $X = Y = Z = 0$, and it follows that (19.1) is insoluble except with $x_1 = \cdots = x_n = 0$. On the other hand it can be proved that (19.1) satisfies the congruence conditions for every $p$, and it is of course also soluble non-singularly in the real field.

Hence some further condition must be imposed if we are to establish solubility in integers. The type of condition which Birch is led to impose is expressed in terms of the dimension of a 'singular locus' associated with the system of equations.

We shall outline the general plan of his paper, giving comparisons with the problem of one cubic equation in places where this may be helpful. The details are somewhat formidable because of the inevitable complexity of the notation.

Suppose we have $R$ homogeneous forms of degree $k$ in $n$ variables, where $R < n$. We can write them as

$$\begin{aligned} f^{(1)}(\mathbf{x}) &= \sum_{j_0,\ldots,j_{k-1}} C^{(1)}_{j_0,\ldots,j_{k-1}} x_{j_0} \ldots x_{j_{k-1}}, \\ &\vdots \\ f^{(R)}(\mathbf{x}) &= \sum_{j_0,\ldots,j_{k-1}} C^{(R)}_{j_0,\ldots,j_{k-1}} x_{j_0} \ldots x_{j_{k-1}}, \end{aligned}$$

where the variables of summation go from 1 to $n$. Let $\mathfrak{B}$ be a box in $n$ dimensional space, and define the exponential sum

$$S(\alpha_1, \ldots, \alpha_R) = \sum_{\mathbf{x} \in P\mathfrak{B}} e\left(\alpha_1 f^{(1)}(\mathbf{x}) + \cdots + \alpha_R f^{(R)}(\mathbf{x})\right).$$

Then the number of integer points $\mathbf{x}$ in $P\mathfrak{B}$ which satisfy the simultaneous equations $f^{(1)}(\mathbf{x}) = 0, \ldots, f^{(R)}(\mathbf{x}) = 0$ is given by

$$\mathcal{N}(P) = \int_0^1 \cdots \int_0^1 S(\alpha_1, \ldots, \alpha_R)\, d\alpha_1 \ldots\, d\alpha_R.$$

By a straightforward extension of Lemma 13.1, we find that if

$$|S(\alpha_1, \ldots, \alpha_R)| \geq P^{n-K}, \quad (K > 0),$$

then

$$\sum_{\mathbf{x}^{(1)}} \cdots \sum_{\mathbf{x}^{(k-1)}} \prod_{J=1}^{n} \min\left\{P, \left\|\alpha_1 M_J^{(1)} + \cdots + \alpha_R M_J^{(R)}\right\|^{-1}\right\} \gg P^{nk-2^{k-1}K},$$

where $M_J^{(1)}, \ldots, M_J^{(R)}$ are the multilinear forms in $k-1$ points $\mathbf{x}^{(1)}, \ldots, \mathbf{x}^{(k-1)}$ defined by

$$M_J^{(i)}\left(\mathbf{x}^{(1)} \mid \cdots \mid \mathbf{x}^{(k-1)}\right) = \sum_{j_1, \ldots, j_{k-1}} c^{(i)}_{J, j_1, \ldots, j_{k-1}} x^{(1)}_{j_1} \cdots x^{(k-1)}_{j_{k-1}},$$

for $i = 1, \ldots, R$. Lemma 13.1 itself is the case $R = 1$, $k = 3$. As in Lemma 13.2, it follows that the number of sets of $k-1$ integer points which satisfy

$$\left|\mathbf{x}^{(1)}\right| < P, \ldots, \left|\mathbf{x}^{(k-1)}\right| < P,$$
$$\left\|\alpha_1 M_J^{(1)} + \cdots + \alpha_R M_J^{(R)}\right\| < P^{-1}, \quad (1 \le J \le n),$$

is

$$\gg P^{(k-1)n - 2^{k-1}K - \varepsilon}.$$

Using Lemma 12.6 $k-1$ times (instead of twice, as in the proof of Lemma 13.3), we deduce that the number of sets of $k-1$ integer points satisfying

$$\left|\mathbf{x}^{(1)}\right| < P^{\theta}, \ldots, \left|\mathbf{x}^{(k-1)}\right| < P^{\theta},$$
$$\left\|\alpha_1 M_J^{(1)} + \cdots + \alpha_R M_J^{(R)}\right\| < P^{-k+(k-1)\theta}$$

is

$$\gg P^{(k-1)n\theta - 2^{k-1}K - \varepsilon}.$$

If there is any one of these sets of $k-1$ points for which the rank of the matrix

$$\begin{pmatrix} M_1^{(1)} & \cdots & M_1^{(R)} \\ \vdots & & \vdots \\ M_n^{(1)} & \cdots & M_n^{(R)} \end{pmatrix}$$

is $R$, then we get good rational approximations to $\alpha_1, \ldots, \alpha_R$, all with the same denominator $q$. This denominator arises as the value of some determinant (non-zero) of order $R$ in the above matrix. In fact we get

$$|q\alpha_i - a_i| \ll P^{-k+R(k-1)\theta}$$

and

$$q \ll P^{R(k-1)\theta}.$$

The exponents here correspond to $-3 + 2\theta$ and $2\theta$ respectively, as in alternative B of Lemma 13.4.

The real difficulty is when the above fails, i.e., when the rank of the above matrix is $\leq R-1$ for all sets $\mathbf{x}^{(1)}, \ldots, \mathbf{x}^{(k-1)}$. In the case $R=1$ this would mean that the multilinear forms $M_J$ all vanish at all these sets of integer points.

The main new idea of Birch's paper is to express this possibility in terms of dimensions of varieties. We regard a set of $k-1$ points as a single point in a space of $(k-1)n$ dimensions. The condition that the rank of the above matrix shall be $\leq R-1$ defines an algebraic variety in this space; and from the lower bound for the number of integer points on it, we deduce that the dimension of this variety is

$$\geq (k-1)n - 2^{k-1}K/\theta + \varepsilon.$$

It is a simple principle of algebraic geometry that the dimension of a variety (that is, the maximum dimension of any of its absolutely irreducible components) cannot be reduced by more than $t$ if we pass to the intersection of the variety with a linear space defined by $t$ equations. Hence the intersection of the above variety with the 'diagonal' linear space

$$\mathbf{x}^{(1)} = \mathbf{x}^{(2)} = \cdots = \mathbf{x}^{(k-1)},$$

defined by $(k-2)n$ equations, has dimension

$$\geq n - 2^{k-1}K/\theta - \varepsilon.$$

If $\mathbf{x} = \mathbf{x}^{(1)} = \cdots = \mathbf{x}^{(k-1)}$, the new variety consists of all points $\mathbf{x}$ for which the rank of the matrix

$$\begin{pmatrix} \frac{\partial f^{(1)}}{\partial x_1} & \cdots & \frac{\partial f^{(R)}}{\partial x_1} \\ \vdots & & \vdots \\ \frac{\partial f^{(1)}}{\partial x_n} & \cdots & \frac{\partial f^{(R)}}{\partial x_n} \end{pmatrix}$$

is $\leq R-1$. We call this *the singular locus associated with the given equations*, and denote it by $V^*$. Thus the present case leads to

$$\dim V^* \geq n - 2^{k-1}K/\theta - \varepsilon.$$

If $\dim V^* = s$, we can prevent this happening (and thereby exclude the situation which now corresponds to alternative A of Lemma 13.4) by choosing

$$K = \frac{\theta}{2^{k-1}}(n - s - 2\varepsilon).$$

Having made this choice, we have a situation similar to that of alternative B; that is, for each $\alpha_1, \ldots, \alpha_R$ there is either an estimate for $|S(\alpha_1, \ldots, \alpha_R)|$ or a good set of simultaneous approximations to $\alpha_1, \ldots, \alpha_R$. This proves the basis for a treatment similar in principle to that of Chapters 15, 16, 17 for one cubic equation.

The main difficulty lies with the singular integral, and here again the dimension of the singular variety comes in. The treatment of the integral is too elaborate to be outlined here. It is essential to suppose that the original equations define a variety of dimension $n - R$.

The result of Birch's paper is as follows:

**Theorem 19.1.** *Let $f_1, \ldots, f_R$ be forms of degree $k$ in $n$ variables with integral coefficients, where $n > R \geq 1$. Let $V$ denote the algebraic variety*

$$f_1(\mathbf{x}) = 0, \ldots, f_R(\mathbf{x}) = 0,$$

*and suppose $V$ has dimension $n - R$. Let $V^*$ be the associated singular locus, and let $s = \dim V^*$. Suppose there is a non-singular real point on $V$, and a non-singular $p$-adic point on $V$ for every prime $p$. Then provided*

$$n - s > R(R+1)(k-1)2^{k-1},$$

*there is an integer point $\mathbf{x} \neq \mathbf{0}$ on $V$.*

# 20

# A Diophantine inequality

In the subject of Diophantine inequalities, our aim is to solve in integers some given type of inequality, and usually one involving polynomials or forms with arbitrary real coefficients. The geometry of numbers provides useful methods for investigating the solubility of linear inequalities, and gives some information about inequalities involving polynomials of higher degree, but is limited in its power in relation to the latter.

The simplest Diophantine inequality of higher degree than the first is

$$|\lambda_1 x_1^2 + \cdots + \lambda_n x_n^2| < C.$$

On the basis of analogy with Meyer's theorem which was encountered in Chapter 11, it was conjectured by Oppenheim in 1929 that provided $n \geq 5$, the inequality should be soluble for *all* $C > 0$, provided that $\lambda_1, \ldots, \lambda_n$ are real numbers which are not all of the same sign. Of course, if $\lambda_1, \ldots, \lambda_n$ are in rational ratios, we can make the left-hand side zero, so the problem relates to the case in which the ratios are not all rational.

In 1934 the result was proved to hold if $n \geq 9$ by Chowla [13]; he deduced it from results of Jarník and Walfisz [51] on the number of integer points in a large ellipsoid. In 1945 it was proved to hold for $n \geq 5$ by Davenport and Heilbronn [26], and the present chapter is mainly devoted to an account of the proof. It should be noted that although the bound 5 for the number of variables is in some sense best possible, there is a deeper sense in which it is probably not. If we assume that the ratios $\lambda_i/\lambda_j$ are not all rational, the result may (for all we know to the contrary) hold for $n \geq 3$.

Stated formally, the result to be proved is:

**Theorem 20.1.** *Let* $\lambda_1, \ldots, \lambda_5$ *be real numbers, none of them* 0, *and* *not all positive nor all negative. Suppose that one at least of the ratios* $\lambda_i/\lambda_j$ *is irrational. Then, for any* $\varepsilon > 0$ *there exist integers* $x_1, \ldots, x_5$, *not all* 0, *such that*

$$|\lambda_1 x_1^2 + \cdots + \lambda_5 x_5^2| < \varepsilon.$$

We can suppose without loss of generality that

$$\lambda_1 > 0, \quad \lambda_5 < 0, \quad \lambda_1/\lambda_2 \notin \mathbb{Q}.$$

It will suffice to prove the solubility of

$$|\lambda_1 x_1^2 + \cdots + \lambda_5 x_5^2| < 1, \tag{20.1}$$

since then the solubility of the apparently more general inequality follows on replacing $\lambda_1, \ldots, \lambda_5$ in the last inequality by $\lambda_1/\varepsilon, \ldots, \lambda_5/\varepsilon$.

The first step is to construct a function of a real variable $Q$ which is positive for $|Q| < 1$ and zero for $|Q| \geq 1$. One such is given in the following lemma, but there are various other similar ones.

**Lemma 20.1.** *We have*

$$\int_{-\infty}^{\infty} e(\alpha Q) \left( \frac{\sin \pi\alpha}{\pi\alpha} \right)^2 d\alpha = \begin{cases} 1 - |Q|, & |Q| \leq 1, \\ 0, & |Q| \geq 1. \end{cases}$$

*Proof.* It is well known that

$$\int_{-\infty}^{\infty} \left( \frac{\sin \pi\alpha}{\pi\alpha} \right)^2 d\alpha = 1.$$

Hence

$$\int_{-\infty}^{\infty} \left( \frac{\sin \pi\eta\alpha}{\pi\alpha} \right)^2 d\alpha = |\eta|$$

for any real $\eta$. This gives

$$\begin{aligned}
&\int_{-\infty}^{\infty} e(\alpha Q) \left( \frac{\sin \pi\alpha}{\pi\alpha} \right)^2 d\alpha \\
&= \int_{-\infty}^{\infty} \cos 2\pi\alpha Q \left( \frac{\sin \pi\alpha}{\pi\alpha} \right)^2 d\alpha \\
&= \frac{1}{2} \int_{-\infty}^{\infty} \frac{\sin^2 \pi\alpha(Q+1) + \sin^2 \pi\alpha(Q-1) - 2\sin^2 \pi\alpha Q}{(\pi\alpha)^2} d\alpha \\
&= \frac{1}{2} \{ |Q+1| + |Q-1| - 2|Q| \},
\end{aligned}$$

which gives the result. □

Let $P$ be a large positive integer. Define

$$S(\alpha) = \sum_{x=1}^{P} e(\alpha x^2), \quad I(\alpha) = \int_0^P e(\alpha x^2)dx.$$

Taking $Q = \lambda_1 x_1^2 + \cdots + \lambda_5 x_5^2$ in the result of Lemma 20.1 and summing over $x_1, \ldots, x_5$ we get

$$\int_{-\infty}^{\infty} S(\lambda_1\alpha)\cdots S(\lambda_5\alpha)\left(\frac{\sin \pi\alpha}{\pi\alpha}\right)^2 d\alpha = \sum_{\substack{x_1,\ldots,x_5 \\ |Q|<1}} (1-|Q|), \quad (20.2)$$

where the summation is over integers with $1 \leq x_j \leq P$ subject to (20.1). Similarly, integrating over $x_1, \ldots, x_5$ instead of summing, we get

$$\int_{-\infty}^{\infty} I(\lambda_1\alpha)\cdots I(\lambda_5\alpha)\left(\frac{\sin \pi\alpha}{\pi\alpha}\right)^2 d\alpha = \int\cdots\int(1-|Q|)dx_1\cdots dx_5, \quad (20.3)$$

where the integration is over real variables with $0 \leq x_j \leq P$ subject to (20.1).

The general idea of the proof is to compare (20.2) with (20.3). It will be an easy matter to prove that the right-hand side of (20.3) is $\gg P^3$ as $P \to \infty$ (see Lemma 20.2 below). If we could prove that the left-hand sides of (20.2) and (20.3) differ by an amount which is $o(P^3)$ as $P \to \infty$, it would follow that the right-hand side of (20.2) was $\gg P^3$. This would imply that there are $\gg P^3$ integral solutions $(x_1, \ldots, x_5)$ of (20.1), with $1 \leq x_j \leq P$.

We shall prove that there is a small interval around $\alpha = 0$ in which $S_j(\alpha)$ differs very little from $I_j(\alpha)$, and from this we shall deduce that the contributions made by this interval to the two integrals are effectively the same (Lemma 20.4). It will be easy to prove that all other $\alpha$ make a negligible contribution to the integral on the left-hand side of (20.3). The difficulty lies in estimating the contribution made by such $\alpha$ to the integral on the left side of (20.2). Here (and here only) we use the hypothesis that $\lambda_1/\lambda_2$ is irrational, and we shall not prove the result in question for *all* large $P$ but only for a particular sequence.

**Lemma 20.2.** *We have*

$$\int_{-\infty}^{\infty} I(\lambda_1\alpha)\cdots I(\lambda_5\alpha)\left(\frac{\sin \pi\alpha}{\pi\alpha}\right)^2 d\alpha \gg P^3.$$

*Proof.* In the right-hand side of (20.3) we put $|\lambda_i|x_i^2 = y_i$. The integral becomes (apart from the constant factor)

$$\int_0^{|\lambda_1|P^2} \cdots \int_0^{|\lambda_5|P^2} \{1 - |y_1 \pm y_2 \pm \cdots - y_5|\} (y_1 \cdots y_5)^{-1/2} dy_1 \cdots dy_5,$$

where the integral is over $y_1, \ldots, y_5$ for which $|y_1 \pm y_2 \pm \cdots - y_5| < 1$ and the signs are those of $\lambda_1, \ldots, \lambda_5$.

We limit the variables $y_2, y_3, y_4$ to the interval $\frac{1}{2}\gamma P^2 < y_j < \gamma P^2$, and we limit $y_5$ to the interval $4\gamma P^2 < y_5 < 5\gamma P^2$, and we limit $y_1$ to the interval

$$|y_1 \pm y_2 \pm y_3 \pm y_4 - y_5| < \frac{1}{2}.$$

Then all the remaining points have $0 < y_j < |\lambda_j|P^2$ provided $9\gamma < \min|\lambda_j|$. Hence we have a portion of the domain of integration, of volume $\gg (P^2)^4$. In this domain, the integrand is

$$\gg (y_1 \cdots y_5)^{-1/2} \gg (P^{10})^{-1/2}.$$

Hence the integral is $\gg P^3$. □

**Lemma 20.3.** *If* $|\alpha| < (4\lambda P)^{-1}$, *where* $\lambda = \max|\lambda_j|$, *then*

$$S(\lambda_j\alpha) = I(\lambda_j\alpha) + O(1).$$

*Proof.* This is a case of Lemma 9.1 (van der Corput's lemma), with $f(x) = \lambda_j\alpha x^2$. We have

$$|f'(x)| = |2\lambda_j\alpha x| \le 2|\lambda_j\alpha|P \le \frac{1}{2},$$

and $f''(x)$ is of fixed sign. □

**Lemma 20.4.** *We have*

$$\int_{|\alpha|<(4\lambda P)^{-1}} S(\lambda_1\alpha)\cdots S(\lambda_5\alpha)\left(\frac{\sin \pi\alpha}{\pi\alpha}\right)^2 d\alpha \gg P^3.$$

*Proof.* First we note that, for any $\alpha$,

$$|I(\lambda_j\alpha)| \ll \min(P, |\alpha|^{-1/2}).$$

The estimate $P$ is obvious, and the estimate $|\alpha|^{-1/2}$ follows from

$$I(\lambda_j\alpha) = \int_0^P e(\lambda_j\alpha x^2)dx = \frac{1}{2}|\lambda_j\alpha|^{-1/2} \int_0^{|\lambda_j\alpha|P^2} t^{-1/2} e(\pm t)dt,$$

since the last integral is bounded. It follows from Lemma 20.3 that if $|\alpha| < (4\lambda P)^{-1}$ then

$$|S(\lambda_j\alpha)| \ll \min(P, |\alpha|^{-1/2})$$

also. Hence, using these two estimates in conjunction with Lemma 20.3, we have

$$|S(\lambda_1\alpha)\cdots S(\lambda_5\alpha) - I(\lambda_1\alpha)\cdots I(\lambda_5\alpha)| \ll \min(P^4, \alpha^{-2})$$

in the above interval. Hence the integral of the difference over $|\alpha| < (4\lambda P)^{-1}$ is $O(P^2)$.

Hence it suffices to prove that

$$\int_{|\alpha|<(4\lambda P)^{-1}} I(\lambda_1\alpha)\cdots I(\lambda_5\alpha)\left(\frac{\sin\pi\alpha}{\pi\alpha}\right)^2 d\alpha \gg P^3.$$

We already know that this is true for the corresponding integral over $(-\infty, \infty)$. Now by the above estimate for $I(\lambda_j\alpha)$, we have

$$\int_{|\alpha|\geq(4\lambda P)^{-1}} |I(\lambda_1\alpha)\cdots I(\lambda_5\alpha)|d\alpha \ll \int_{|\alpha|\geq(4\lambda P)^{-1}} \alpha^{-5/2}d\alpha \ll P^{3/2}.$$

Hence the result is proved. □

We now come to the heart of the problem; that is, the estimation of

$$\int_{|\alpha|\geq(4\lambda P)^{-1}} |S(\lambda_1\alpha)\cdots S(\lambda_5\alpha)|\left(\frac{\sin\pi\alpha}{\pi\alpha}\right)^2 d\alpha.$$

**Lemma 20.5.** *For any $\varepsilon > 0$ we have*

$$\int_0^1 |S(\alpha)|^4 d\alpha \ll P^{2+\varepsilon}.$$

*Proof.* By the definition of $S(\alpha)$, the integral equals the number of solutions of $x_1^2 + x_2^2 = y_1^2 + y_2^2$ in integers between 1 and $P$ inclusive. The number of solutions with $x_2 = y_2$ is $P^2$. In other solutions, the value of $x_2$ and $y_2$ determine those of $x_1 - y_1$ and $x_1 + y_1$ with $\ll P^\varepsilon$ possibilities, since these are factors of $x_2^2 - y_2^2$. Hence the result. □

**Lemma 20.6.** *For any fixed $\delta > 0$ we have*

$$\int_{|\alpha|\geq P^\delta} |S(\lambda_1\alpha)\cdots S(\lambda_5\alpha)|\left(\frac{\sin\pi\alpha}{\pi\alpha}\right)^2 d\alpha \ll P^{3-\delta/2}.$$

*Proof.* In view of the trivial estimate $|S(\lambda_5\alpha)| \leq P$, it suffices to prove that

$$\int_{|\alpha|\geq P^\delta} |S(\lambda_1\alpha)\cdots S(\lambda_4\alpha)| \left(\frac{\sin \pi\alpha}{\pi\alpha}\right)^2 d\alpha \ll P^{2-\delta/2}.$$

By Hölder's inequality, it suffices to prove that

$$\int_{|\alpha|\geq P^\delta} |S(\lambda_j\alpha)|^4 \left(\frac{\sin \pi\alpha}{\pi\alpha}\right)^2 d\alpha \ll P^{2-\delta/2},$$

and hence it suffices to prove that

$$\int_{|\alpha|\geq P^\delta} |S(\alpha)|^4 \frac{d\alpha}{\alpha^2} \ll P^{2-\delta/2}.$$

Since $S(\alpha)$ is periodic with period 1, Lemma 20.5 implies that

$$\int_m^{m+1} |S(\alpha)|^4 d\alpha \ll P^{2+\varepsilon}.$$

Hence, if $M = [P^\delta]$, the integral in question is at most

$$\begin{aligned} \sum_{m=M}^{\infty} \int_m^{m+1} |S(\alpha)|^4 \frac{d\alpha}{\alpha^2} &\ll P^{2+\varepsilon} \sum_{m=M}^{\infty} \int_m^{m+1} \frac{d\alpha}{\alpha^2} \\ &\ll P^{2+\varepsilon} M^{-1} \\ &\ll P^{2-\delta+\varepsilon} \\ &\ll P^{2-\delta/2}. \end{aligned}$$

Hence the result. □

By (20.2) and Lemma 20.4 and Lemma 20.6, it suffices to prove that

$$\int_{(4\lambda P)^{-1}<|\alpha|<P^\delta} |S(\lambda_1\alpha)\cdots S(\lambda_5\alpha)| d\alpha = o(P^3).$$

For then the right-hand side of (20.2) is $\gg P^3$, and this is what we want to prove. As mentioned earlier, we can only prove this for certain restricted values of $P$.

It follows from Lemma 20.5 and Hölder's inequality that

$$\int_{(4\lambda P)^{-1}<|\alpha|<P^\delta} |S(\lambda_{i_1}\alpha)\cdots S(\lambda_{i_4}\alpha)| d\alpha \ll P^{2+\delta+\varepsilon}$$

for any four different subscripts $i_1, \ldots, i_4$. Using this with the subscripts $2, 3, 4, 5$ and $1, 3, 4, 5$, we see that it will suffice if, for each $\alpha$ in the range of integration, we have

$$\min\left(|S(\lambda_1\alpha)|, |S(\lambda_2\alpha)|\right) \ll P^{1-2\delta}. \tag{20.4}$$

For this we must use the irrationality of $\lambda_1/\lambda_2$.

**Lemma 20.7.** *Suppose that* $(a,q)=1$ *and* $|\alpha - a/q| < q^{-2}$. *Then*

$$|S(\alpha)| \ll P^{1+\varepsilon}\left\{P^{-1/2} + q^{-1/2} + \left(\frac{P^2}{q}\right)^{-1/2}\right\}.$$

*Proof.* This is Lemma 3.1 (Weyl's inequality), with $k=2$, and consequently also $K=2$. □

We choose any convergent $a_0/q_0$ to the continued fraction for $\lambda_1/\lambda_2$, and therefore have

$$\left|\frac{\lambda_1}{\lambda_2} - \frac{a_0}{q_0}\right| < \frac{1}{q_0^2}. \tag{20.5}$$

We take $P = q_0^2$; this limits $P$ to an infinite sequence of values.

**Lemma 20.8.** *With* $P$ *restricted as above, the estimate (20.4) holds for each* $\alpha$ *in the range* $(4\lambda P)^{-1} < \alpha < P^{\delta}$.

*Proof.* There exists a rational approximation $a_1/q_1$ to $\lambda_1\alpha$ such that

$$(a_1, q_1) = 1, \quad 1 \le q_1 \le P^{3/2}, \quad \left|\lambda_1\alpha - \frac{a_1}{q_1}\right| < \frac{1}{q_1 P^{3/2}}. \tag{20.6}$$

We observe that $a_1 \neq 0$, for if $a_1 = 0$ then $|\lambda_1\alpha| < P^{-3/2}$, contrary to hypothesis. Similarly there exists a rational approximation $a_2/q_2$ to $\lambda_2\alpha$ such that

$$(a_2, q_2) = 1, \quad 1 \le q_2 \le P^{3/2}, \quad \left|\lambda_2\alpha - \frac{a_2}{q_2}\right| < \frac{1}{q_2 P^{3/2}}, \tag{20.7}$$

and again $a_2 \neq 0$.

If $q_1 > P^{5\delta}$, we can apply Lemma 20.7 to $S(\lambda_1\alpha)$, with $a_1, q_1$ in place of $a, q$, and this gives[1]

$$|S(\lambda_1\alpha)| \ll P^{1+\varepsilon-5\delta/2} \ll P^{1-2\delta}.$$

Similarly if $q_2 > P^{5\delta}$ we get the analogous result for $|S(\lambda_2\alpha)|$. In either of these events, (20.4) is satisfied. So we can suppose that

$$q_1 \le P^{5\delta}, \quad q_2 \le P^{5\delta}. \tag{20.8}$$

[1] We assume that $\delta$ is small.

We can now deduce from (20.6) and (20.7) that $\lambda_1/\lambda_2$ is well approximated by $a_1q_2/a_2q_1$. We note that since $a_1/q_1$ is an approximation to $\lambda_1\alpha$, and $|\alpha| < P^\delta$, we have $|a_1| \ll P^{6\delta}$, and similarly $|a_2| \ll P^{6\delta}$. Hence

$$\frac{\lambda_1}{\lambda_2} = \frac{\lambda_1\alpha}{\lambda_2\alpha} = \frac{\frac{a_1}{q_1}(1+O(P^{-3/2}))}{\frac{a_2}{q_2}(1+O(P^{-3/2}))} = \frac{a_1q_2}{a_2q_1}\left(1+O(P^{-3/2})\right),$$

and since $|a_1q_2/a_2q_1|$ is bounded above, this implies

$$\left|\frac{\lambda_1}{\lambda_2} - \frac{a_1q_2}{a_2q_1}\right| \ll P^{-3/2}.$$

We have $1 \le |a_2|q_1 \ll P^{11\delta}$.

Comparison of the last result with (20.5) gives a contradiction if $\delta$ is sufficiently small (and $q_0$ is sufficiently large). For we get

$$\left|\frac{\lambda_1}{\lambda_2} - \frac{a_1q_2}{a_2q_1}\right| \ll P^{-3/2} + q_0^{-2} \ll q_0^{-2},$$

for $P = q_0^2$, whereas the left-hand side is

$$\ge \frac{1}{q_0|a_2|q_1} \gg \frac{1}{q_0P^{11\delta}} \gg q_0^{-1-6\delta}.$$

This completes the proof of Lemma 20.8; and by our earlier remarks, Lemma 20.8 completes the proof of Theorem 20.1. □

Certain extensions of Theorem 20.1 are almost immediate. First, we could prove the same result for the inequality

$$|\lambda_1x_1^2 + \cdots + \lambda_5x_5^2 - \mu| < \varepsilon,$$

for any real number $\mu$ (assuming that the ratios $\lambda_i/\lambda_j$ are not all rational). Secondly we could replace the squares by $k$th powers, provided the number of variables is at least $2^k+1$. In this case we should use Hua's inequality (Lemma 3.2) in place of the above Lemma 20.7. More precise results have been proved by Davenport and Roth [30] and by Danicic [16].

The extension to Diophantine inequalities involving general forms with real coefficients, in place of additive forms, present perhaps even more difficulty than the corresponding extension for Diophantine equations.

All the results so far obtained depend on results for Diophantine equations, and usually one needs these in a more precise form, in which there is an estimate for the size of a solution.

The first problem that naturally presents itself is that of establishing the solubility of

$$|Q(x_1, \ldots, x_n)| < \varepsilon \tag{20.9}$$

for any $\varepsilon > 0$, where $Q$ is any indefinite quadratic form. By some very complicated work, this has been proved to hold if $n \geq 21$, the work being the result of joint efforts by Birch, Davenport and Ridout [70]. By a result of Oppenheim [64, 65], it follows that if $Q$ is not proportional to a form with integral coefficients, then the inequality

$$|Q(x_1, \ldots, x_n) - \mu| < \varepsilon,$$

for any real $\mu$, is soluble. Thus the values of any real indefinite quadratic form in 21 or more variables are either discrete (if the form is proportional to an integral form) or everywhere dense.

An analogue of (20.9) for any real cubic form has been proved by Pitman [66], but the number of variables needed is fairly large. There seems to be a difficulty of principle in proving any analogous result for a form of degree five.

# Index

Printed in the United States
By Bookmasters